Rules, Norms and NGO Advocacy Strategies

There is much controversy over the development of new dams for hydropower, where concerns for environmental protection and the livelihoods of local people may conflict with the goals of economic development. This book analyses the opportunities and barriers that NGOs and civil society actors face when conducting advocacy campaigns against such developments.

Through a comparison of two NGO coalitions advocating against the Xayaburi hydropower dam on the Mekong River, the book explores the intricate interactions of formal and informal rules and norms, and how they influence advocacy strategies. A framework for analysis is proposed which serves as a tool for use by civil society actors. The author generates fresh insights into the creation of opportunities – and barriers – for NGOs aiming to influence state-centric decision-making processes.

The book also discusses in detail Mekong riparian states' negotiation process over the Xayaburi hydropower dam, providing an analysis of the Mekong River's governance under the 1995 Mekong Agreement. It concludes by suggesting ways to improve the engagement of civil society actors in the governance of transboundary rivers and development projects.

Yumiko Yasuda is an environmental and water governance specialist. She was a researcher for the Centre for Water Law, Policy and Science, University of Dundee, Scotland, UK. She has previously worked for the World Wide Fund for Nature (WWF) and the United Nations Development Programme (UNDP) in Southeast Asia.

Rules, Norms and NGO Advocacy Strategies

Hydropower development on the Mekong River

Yumiko Yasuda

Routledge
Taylor & Francis Group

LONDON AND NEW YORK

earthscan
from Routledge

First published 2015 by Routledge

2 Park Square, Milton Park, Abingdon, Oxfordshire OX14 4RN

711 Third Avenue, New York, NY 10017

Routledge is an imprint of the Taylor & Francis Group, an informa business

First issued in paperback 2017

© 2015 Yumiko Yasuda

British Library Cataloguing in Publication Data
A catalogue record for this book is available from the British Library

Library of Congress Cataloging in Publication Data
A catalog record has been applied for

ISBN: 978-1-138-92029-3 (hbk)
ISBN: 978-0-8153-9537-9 (pbk)

Typeset in Bembo
by Saxon Graphics Ltd, Derby

Contents

Figures

Tables

Acknowledgements

The journey to complete this book was one of the most rewarding experiences of my life. I was fortunate to be accompanied by so many people who supported me during this journey.

I would like to thank Mr Tim Hardwick and Ms Ashley Wright from Routledge/Taylor & Francis Group for providing me with the opportunity to write this book, and thank editorial staff for their valuable feedback and support they provided. I am grateful to Professor Geoffrey Gooch and Dr Alistair Rieu-Clarke for their support and patience through engaging in hours of discussions with me on the subject, and providing constructive criticisms on my writings. Many thanks to Dr. Jamie Pittock for encouraging me to publish my work, and Dr. Janet Liao and Dr. Sarah Hendry for providing constructive feedback on my work. Mr Sopheak Chanh from the University of Sydney provided me with a map used in this book. Many thanks to Stefan and Namiko for providing me with creative ideas about the book.

I would like to thank all the interviewees who made themselves available, and told me so many interesting stories. Special thanks to members of the Vietnam Rivers Network and the Rivers Coalition in Cambodia, for being supportive of my research idea and being so collaborative. I sincerely hope my research will be useful to all of you. I am grateful to colleagues at the Worldfish Centre Phnom Penh Office for kindly accommodating me during my stay in Phnom Penh. My book also benefited from discussions and feedback from many other people I encountered: feedback I received during conferences, conversations I had with friends while sitting by the Mekong, casual chats with villagers in Cambodia during my fieldwork, and many other occasions. Although I cannot mention each name, I would like to thank every one of you who gave me inspiration and encouragement throughout my journey.

Many friends and colleagues, who live all over the world, supported my work. My friends and colleagues at University Dundee always remained by my side, during the time of difficulties, sharing tears and laughter. I would like to thank Amanda, Graham, Satomi, Aya, Junichi, Liana, Peter-John, Nancy and JD for generously accommodating me during my fieldwork, and introducing me to their local contacts. My friends Hoang and Hung helped me to understand some of the Vietnamese documents. I would like to thank Robert for always sharing valuable information and his insights from the Mekong. Many thanks to Dr. Philip Andrews-Speed and Erin for their feedback on my draft chapters. Academic discussions I had with researchers at the Mekong Research Group at the University of Sydney were valuable for analysing my field data. Special thanks to Professor Philip Hirsch for his kind support.

The research for the book was funded by the Joint Japan/World Bank Graduate Scholarship Programme and the Endeavour Research fellowship. Travel to Vietnam through the LiveDiverse project meeting gave me the opportunity to conduct my initial fieldwork in November 2011.

Finally, I would like to thank my family for always being supportive of my challenge in life. Without their unconditional love and encouragement, my journey would not have been completed.

Abbreviations

3SPN	3S Rivers Protection Network
ADB	Asian Development Bank
AMRC	Australian Mekong Resource Centre
ANT	actor network theory
ASEAN	Association of Southeast Asian Nations
BAED	Buddhist Association for Environmental Development
B&M	biophysical and material (condition)
BOT	build-operate-transfer
CAN	Climate Action Network
CARD	Council for Agriculture and Rural Development
CBD	Centre for Biodiversity and Development
CBO	community-based organization
CCC	Cooperation Committee for Cambodia
CCIM	Cambodian Center for Independent Media
CED	Community Economic Development
CEPA	Culture and Environment Preservation Association
CIVICUS	World Alliance for Citizen Participation
CNMC	Cambodian National Mekong Committee
CNR	Compagnie Nationale du Rhône
COP	Conference of Parties
CPP	Cambodian People's Party
CPV	Communist Party of Vietnam
CRDT	Cambodian Rural Development Team
CSO	civil society organization
CSRD	Centre for Social Research and Development
ECAFE	United Nations Economic Commission for Asia and the Far East
Eco-Eco	Institute of Ecological Economy
EGAT	Electricity Generating Authority of Thailand
EIA	environmental impact assessment
ERU	equitable and reasonable utilization
ESCR	United Nations Committee on Economic and Social and Cultural Rights

EU	European Union
EVN	Electricity Vietnam
FACT	Fisheries Action Coalition Team
ForWet	Research Centre of Forest and Wetlands
FPIC	free prior and informed consent
FUNCINPEC	Front uni national pour un Cambodge indépendant, neutre, pacifique, et coopératif [National United Front for an Independent, Neutral, Peaceful, and Cooperative Cambodia]
GMS	Greater Mekong Sub-region
GONGO	government-organized NGO
IAD framework	institutional analysis and development framework
ICCO	Internationaal Maatschappelijk Verantwoord Ondernemen [Interchurch Organization for Development Cooperation] (the Netherlands)
ICEM	International Centre for Environmental Management
IR	International Rivers
ISH	Initiative on Sustainable Hydropower
IUCN	International Union for Conservation of Nature
JC	Joint Committee
LANGO	Law on Associations and Non-Governmental Organizations (Cambodia)
Laos	Lao People's Democratic Republic
LMB	Lower Mekong Basin
LNMC	Lao National Mekong Committee
MLN	Mekong Legal Network
MoI	Ministry of the Interior (Cambodia)
MONRE	Ministry of Natural Resources and Environment (Vietnam)
MoU	Memorandum of Understanding
MOWRAM	Ministry of Water Resources and Meteorology (Cambodia)
MRC	Mekong River Commission
MW	megawatt
MWBP	Mekong Wetlands Biodiversity Conservation and Sustainable Use Programme
NGO	nongovernmental organization
NIE	new institutional economics
NMC	National Mekong Committee
OECD	Organisation for Economic Co-operation and Development
PNPCA	Procedures for Notification, Prior Consultation and Agreement
PRK	People's Republic of Kampuchea
PWG	Party Working Groups (of the CPP)

QUANGO	quasi-autonomous NGO
RCC	Rivers Coalition in Cambodia
SEA	strategic environmental assessment
SIA	social impact assessment
STM Coalition	Save the Mekong Coalition
TAN	transnational advocacy network
TERRA	Towards Ecological Recovery and Regional Alliance
TNMC	Thai National Mekong Committee
ToR	Terms of Reference
UNCED	United Nations Conference on Environment and Development
UNDP	United Nations Development Programme
UNECE	United Nations Economic Commission for Europe
UNESCAP	United Nations Economic and Social Commission for Asia and the Pacific
UNFCCC	United Nations Framework Convention on Climate Change
UNTAC	United Nations Transitional Authority in Cambodia
VFF	Vietnam Fatherland Front
VNGO	Vietnamese NGO
VNMC	Vietnam National Mekong Committee
VNWP	Vietnam Water Partnership
VRN	Vietnam Rivers Network
VUSTA	Vietnam Union of Science and Technology Associations
WARECOD	Centre for Water Resources Conservation and Development
WCD	World Commission on Dams
WWF	World Wide Fund for Nature

Legal instruments cited

This list is divided into international and national legal instruments. For legal instruments with long names, short names are provided in parentheses; the book refers to the short name in citations.

International instruments

1969

Vienna Convention on the Law of Treaties. Adopted on 23 May 1969, entered into force on 27 January 1980. 1155 United Nations Treaty Series 331. (Vienna Convention)

1973

Convention on International Trade in Endangered Species of Wild fauna and Flora. Adopted on 3 March 1973, entered into force on 1 July 1975. (CITES)

1975

Joint Declaration of Principles for Utilization of the Waters of the Lower Mekong Basin, signed by the representatives of the governments of Cambodia, Laos, Thailand and Vietnam to the Committee for Coordination of Investigations of the Lower Mekong Basin. Signed at Vientiane on 31 January 1975. (Mekong Joint Declaration)

1991

Convention on Environmental Impact Assessment in a Transboundary Context. Adopted on 25 February 1991, entered into force on 10 September 1997. (Espoo Convention)

1995

Agreement on the Cooperation for the Sustainable Development of the Mekong River Basin. 5 April 1995. (Mekong Agreement)

1997

Convention on the Law of Non-Navigational Uses of International Watercourses. Adopted on 21 May 1997, entered into force on 17 August 2014. Reprinted in (1997) 36 ILM 700. (UN Watercourses Convention)

1998

The UNECE Convention on Access to Information, Public Participation in Decision-making and Access to Justice in Environmental Matters. Adopted on 25 June 1998, entered into force on 30 October 2001. 2161 UNTS 447; (1999) 38 ILM 517. (Aarhus Convention)

2003

Procedure for Notification, Prior Consultation and Agreement. Approved on 30 November 2003 by the 10th meeting of the MRC Council. (PNPCA)

2005

Guidelines on Implementation of the Procedures for Notification, Prior Consultation and Agreement. Adopted on 31 August 2005 by the 22nd meeting of the MRC Joint Committee in Vientiane, Lao PDR. (PNPCA Guideline)

National instruments (Cambodia)

1992

Provisions Relating to the Judiciary and Procedure Applicable in Cambodia during the Transitional Period. Decision adopted by the Supreme National Council of Cambodia on 10 September 1992. (UNTAC Criminal Code)

1993

The Constitution of the Kingdom of Cambodia. 1993. (Constitution of Cambodia)

1995

Law on the Press. 1995.

2001

Law on Commune/Sangkat Administrative Management. 2001. (Law on Commune Management)

2005

Law on Senate Election. 2005.

2006

Amended Law on Elections of Commune Councils. 2006.

2007

Law on Water Resources Management of the Kingdom of Cambodia. 2007. (Law on Water Resources Management)

2008

Law on Elections of Capital Council, Provincial Council, Municipal Council, District Council and Khan Council. 2008. (Law on Sub-National Council Election)

The Civil Code of Cambodia. 2008.

2009

Penal Code. 2009. (Penal Code, Cambodia)

2011

Law on Associations and Non-Governmental Organizations (3rd Draft). 2011. (LANGO 3rd Draft)

National instruments (Vietnam)

1992

Constitution of the Socialist Republic of Vietnam: As Amended on 25 December 2001. (Constitution of Vietnam)

Decree 35/HTBT of 28 January 1992 of the Council of Ministers on Establishment of non-profit and science and technology organization. (Decree No. 35-HDBT)

1995

Decision No. 860-TTg of 30 December 1995 of the Prime Minister on the Function, Tasks, Powers and Organization of the Apparatus of the Vietnam Mekong River Committee. (Decision No. 860-TTg)

1999

Law on Vietnam Fatherland Front. No. 14/1999/QH10 of June 12 1999. The National Assembly.

Penal Code. No. 15/1999/QH10 of 21 December 1999. The National Assembly. (Penal Code, Vietnam)

Law on Media. No. 12/1999/QH10 of 12 June 1999. The National Assembly.

2001

Law on Organization of the Government. No. 32/2001/QH10 of 25 December 2001. The National Assembly.

2002

Decree No. 81/2002/ND-CP of 17 October 2002 of the Government on detailing the implementation of a number of articles of the science and technology law. (Decree No. 81/2002/ND-CP)

Decision 22/2002/QD-TTg of 30 January 2002 of the Prime Minister on the activities of consultancy, judgment and social expertise by Vietnam Union of Scientific and Technical Associations. (Decision No. 22/2002/QD-TTg)

2003

Decree No. 88/2003/ND-CP of 30 July 2003 of the Government on providing for the organization, operation and management of associations. (Decree No. 88/2003/ND-CP)

2006

Regulations of Vietnam Union of Science and Technology Associations. Issued jointly with the decision No. 650/QD-TTg, 24 April 2006 of the Prime Minister of the Socialist Republic of Vietnam "Approval the Regulations of Vietnam Union of S-T Associations." (VUSTA Regulation)

2009

Regulation on management and use of foreign nongovernmental aid promulgated by Decree No. 93/2009/ND-CP of 22 October 2009 of the Government. (Regulation on the use of aid from INGO)

2012

Decree No. 30/2012/ND-CP of 12 April 2012 of the Government on the organization and operation of social funds and charity funds. (Decree No. 30/2012/ND-CP)

Charter. The Vietnam Union of Science and Technology Associations. (VUSTA Charter)

National instrument (other jurisdiction)

2005

Charities and Trustee Investment (Scotland) Act. 2005.

1 NGOs and environmental governance

Role of NGOs in transboundary water governance

Nongovernmental organizations (NGOs) have increasingly become important agents in environmental governance (Florini and Simmons 2000; Betsill and Corell 2008). The engagement of NGOs in this field escalated particularly after the United Nations Conference on the Human Environment in 1972 (Betsill and Corell 2008: 1). Agenda 21, the milestone policy document adopted by the United Nations Conference on Environment and Development (UNCED) in 1992 formally recognized NGOs as important partners in the move to sustainable development (United Nations 1992: ch 7). The role of NGOs can be seen at different administrative and geographical levels, including local, national and international levels (Keck and Sikkink 1999; Florini and Simmons 2000).

NGOs often represent actors dependent on the direct use of natural resources; these actors are often impacted by alterations of resource use through development activities (Bruch 2005: 23). Their engagement in decision-making processes over the use of environmental and natural resources could potentially benefit the subsequent management and governance of such resources (Bruch 2005: 23). Decision-making without consideration of these actors' views, often represented through an NGO, could result in negative consequences when decision-makers face protests or public scrutiny over their decisions (Bruch 2005: 24). NGOs therefore constitute an essential aspect of the governance of the environment and natural resources, and understanding their strategies and behaviours enriches our knowledge and understanding in the field of governance.

In the decision-making process over the use of transboundary rivers, NGOs are often involved in a wide range of advocacy activities associated with hydropower dams. As seen in the case of hydropower dam development around the world, NGOs from the North to the South often create international coalitions in order to raise the voices of local communities to decision-makers (Fischer 1995). As an example, an international campaign to stop Sardar Sarovar–Narmada dam pressured the World Bank to withdraw from the project in India (Fu 2009). Another international coalition of NGOs protested against the construction of 12 hydropower dams in Borneo, Malaysia (Borneo Project 2012). In order to stop the construction of the Xayaburi dam in Lao People's Democratic Republic (Laos), 263 NGOs from 51 countries signed a letter addressed to the Prime Ministers of Thailand and Laos (263 NGOs 2011).

The presence of NGOs and other interest groups could potentially create and enhance either a pluralist or a corporatist society (Almond *et al.* 2004: 69), allowing wider groups in society to engage in water governance. This could gradually facilitate a move away from an elitist model of governance to a more participatory style. NGOs often express the interests of people who are not well represented in the policy-making arena (Charnovitz 1997: 274), playing a role as 'bottom-up brokers'. Ebbesson (2007) claims that NGOs are often granted special rights to represent civil society, as opposed to individual actors, networks and social movements that are generally disregarded (Ebbesson 2007: 689). NGOs and civil society actors are therefore often considered as essential pillars in promoting transparency, accountability and other aspects of 'good governance' (Edwards 2004: 15). Advocacy through civil society is a potential way of promoting policy change towards democratic governance (Fox and Helweg 1997: 8).

NGOs often provide technical information related to the subject of debate as a way of providing input to the policy-making process. Many NGOs are capable of delivering technical expertise on particular topics that government officials need. Government officials can also benefit from this technical expertise through receiving rapid feedback from NGOs on controversial ideas (Charnovitz 1997: 274). NGOs with technical expertise are also capable of contributing to the scientific assessments needed by policy-makers, such as the Millennium Ecosystem Assessment, a process initiated by the United Nations Secretary-General in order to assess the consequences of ecosystem changes for human well-being (Millennium Ecosystem Assessment 2005; see also Gemmill and Bamidei-Izu 2002: 11). This role of intelligence provider also supports improved environmental governance as politicians and decision-makers may gain alternative ideas from NGOs, which would not have been available through normal bureaucratic channels (Charnovitz 1997: 274).

NGOs as promoters of rules and norms

NGOs can also play an important role in promoting the rules and norms associated with environmental governance. Some environmental NGOs are allowed to be present at negotiations on global environmental agreements, where state representatives are the main actors. While NGOs do not have the official negotiating power that state delegates have, NGOs can play important complementary roles to state delegations, since NGOs represent important civil society groups and can attempt to integrate their proposed norms into the negotiating table (Betsill and Corell 2008: 3). A large number of NGOs gather at Conferences of Parties (COP) of multilateral environmental agreements, and may hold side events promoting various issues associated with the themes of COPs (UNFCCC 2011). At times, these NGOs also network with each other to join voices in a common cause. For example, during the negotiations of the Kyoto Protocol under the United Nations Framework Convention on Climate Change (UNFCCC), over 280 NGOs created the Climate Action Network

(CAN) in order to coordinate their access and participation in the COPs (Betsill 2008: 46).

Monitoring states' compliance with law and regulations is another role that NGOs may play (Charnovitz 1997: 274; 2006: 354; Gemmill and Bamidei-Izu 2002). NGOs often raise awareness of state and company violations of legal requirements through filing a complaint or making a claim to a court or authority. If the procedure within a country does not allow an NGO to directly file a complaint, the NGO may provide advice to an individual or an organization which has a right to do so (Bombay 2001: 229). Some NGOs were set up specifically to undertake this task of monitoring compliance with international agreements (Gemmill and Bamidei-Izu 2002: 12). For example, originally established as a specialist group within the International Union for Conservation of Nature (IUCN), TRAFFIC is an NGO with particular focus on monitoring and taking action against wildlife trade restricted under the Convention on International Trade in Endangered Species of Wild Fauna and Flora (CITES) (TRAFFIC 2008). Also, the United Nations Economic Commission for Europe (UNECE) Convention on Access to Information, Public Participation in Decision-making and Access to Justice in Environmental Matters (Aarhus Convention) Article 15 stipulates a compliance committee which monitors states' compliance to the Convention (Article 15 Aarhus Convention 1998). NGOs participate in this compliance committee. Monitoring of legal compliance can be a resource-intensive activity, and engaging NGOs that voluntarily monitor compliance can positively result in saving public resources.

NGOs also play roles as norm entrepreneurs and norm developers (Finnemore and Sikkink 1998). Norms play important roles in a society due to their functions of creating a common understanding of issues, as well as facilitating the emergence, adoption and implementation of new policies and laws (Brown *et al.* 2000: 20). Referring to NGOs' role as norm entrepreneurs, Brown *et al.* (2000) explain that NGOs are often in close contact with vulnerable populations with less voice within a society (Brown *et al.* 2000: 20). Norm entrepreneurs are groups or individuals who are able to influence the acceptance, rejection or development of specific norms (Finnemore and Sikkink 1998). Koh (1998) maintains that NGOs play an important role as norm entrepreneurs within a process of development of norms. Finnemore and Sikkink (1998) discuss the importance of the framing and interpretation of norms by norm entrepreneurs, influencing the acceptance of norms by society (Finnemore and Sikkink 1998: 896–897). Referring to the roles of NGOs as norm entrepreneurs, Finnemore and Sikkink also suggest that the promotion of norms often requires some type of organizational platform, particularly at international levels. NGO networks can play key roles as they are often part of such platforms (Finnemore and Sikkink 1998: 899).

Influence of rules and norms on NGOs

While NGOs attempt to influence rules and norms, rules and norms also affect NGOs. For example, the 1998 Aarhus Convention provides a framework for public right of access to information and participation in the decision-making process. Certain countries establish official mechanisms for public participation, such as public complaint processes or inspection panels. The establishment of citizen enforcement actions allows citizens to take legal action to enforce environmental laws (Casey-Lefkowitz *et al.* 2005: 568; Hunter 2005: 637). These rights may be formulated either as a constitutional right, under specific provisions in environmental law, or in accordance with administrative or civil codes (Casey-Lefkowitz *et al.* 2005: 568; Hunter 2005: 637). A more collaborative approach in some countries obliges government institutions to engage citizens in monitoring their environmental performances (Casey-Lefkowitz *et al.* 2005: 566). As an example, in the USA some citizen organizations initiated harbour watch programmes to identify oil spills and other emissions in local harbours (Casey-Lefkowitz *et al.* 2005: 566). Finally, citizen inspection is a method in some countries whereby government agencies contract with groups or individuals to serve as public inspectors (Casey-Lefkowitz *et al.* 2005: 567).

In contrast, other rules and norms restrict NGOs' activities. As an example, human rights NGOs in Uganda conduct self-censorship when working with sensitive issues, in order to avoid confrontation with the authorities (Ghosh 2009: 483). Ho (2007) suggests that for environmental activism to survive and gain legitimacy in the Chinese restrictive context, it has to be structurally embedded through having 'self-imposed censorship and a conscious de-politicization of environmental politics' (Ho 2007: 189).

These examples indicate that rules and norms affect the way NGOs formulate certain actions and strategies. Thus understanding the influence of rules and norms on NGO actors provides important insights into how NGO actors can contribute to promoting participatory governance. However, compared with the large number of existing studies discussing the ways NGOs attempt to influence rules and norms, studies attempting to understand the influence of rules and norms on NGOs are few and far between. A better understanding of this influence will help fill a significant gap in existing academic knowledge; and improved understanding of these mechanisms will provide useful insights for NGOs themselves, leading to enhanced understanding of the linkages between their actions and the rules and norms within which they operate. Ultimately this may lead to improved governance of environmental and natural resources.

Objective and structure of this book

Recognizing the importance of understanding how rules and norms can potentially influence the activities of NGOs, this book aims to examine and analyse how this influence occurs. Through this understanding, the book identifies the opportunities and barriers NGO actors are facing in their activities. In particular, it examines how rules and norms affect the advocacy strategies of NGO coalitions.

NGOs often form coalitions in order to improve their effectiveness in the process of their advocacy work. For the purpose of this book, coalitions are defined as networks of NGOs and other civil society actors sharing broader strategic goals, beyond a single issue and with the ambition to conduct joint actions. The focus on advocacy strategies was chosen as NGO actors focusing on advocacy tend to be in conflict with the authorities, compared with service sector NGOs, who tend to collaborate with the authorities (Ohanyan 2012). Advocacy NGOs are important actors in governance as their focus is to advocate policy change, which may affect resource use.

The book also has an ambition to contribute to NGO actions that are more informed about the rules and norms affecting them. In order to achieve these objectives, the book aims to understand how rules and norms influence the advocacy strategies of NGO coalitions.

This book consists of three parts. Part I introduces the way of analysing advocacy strategies of NGOs, and consists of two chapters. The first chapter in this section (Chapter 2) provides a theoretical overview and analysis related to analysing NGO advocacy strategies, and the influence of rules and norms over advocacy strategies. This overview and the analysis are used to identify a framework for analysing advocacy strategies of NGO coalitions, discussed in the second chapter of this section (chapter 3). The book adopts a comparative case study approach in analysing the advocacy strategies of two NGO coalitions in Vietnam and Cambodia in Southeast Asia. The focus of these NGO strategies is on the Xayaburi hydropower dam, located on the Mekong River in Laos.

The analytical framework developed in Chapter 3 provides the structure for the book's analysis. This framework identifies five key components important for the analysis of the book: (1) biophysical and material conditions; (2) actors; (3) rules and norms; (4) interactions; (5) strategies. These components guide the structure of the rest of the book.

Part II analyses the first three components: biophysical and material conditions (Chapter 4), rules and norms (Chapter 5), and actors (Chapter 6). Analyses of these three components then feed into analyses of the remaining two components, in order to identify how rules and norms affect the strategies of NGO coalitions.

Part III offers an analysis of how the three components interact and how strategies are developed. Strategies adopted by NGO coalitions during the consultation process of the Xayaburi hydropower dam are first identified in Chapter 7, where four types of strategy depending on the target audiences of advocacy have been identified. Subsequent chapters (8–11) compare strategies adopted by two different NGO coalitions, and how the first three components identified in Part II interact in determining these strategies: Chapter 8 analyses strategies targeting Mekong regional decision-makers; Chapter 9, strategies targeting national decision-makers; Chapter 10, strategies targeting stakeholders in potentially affected areas; and chapter 11, strategies targeting the general public. The book's conclusions are presented in Chapter 12.

The Mekong River, the subject of this book's case studies, faces rapid development pressure. As of 2010, 71 hydropower dam projects were estimated to be in operation on the tributaries of the Mekong by 2030, and 12 hydropower dam projects were planned on the mainstream of the Mekong River (ICEM 2010). Among all the dams on the mainstream of the Lower Mekong River Basin, the Xayaburi hydropower dam is the first one to have undergone a process of prior consultation. The prior consultation process of Don Sahong dam, which is the second hydropower dam project located at the mainstream of the Lower Mekong River Basin, commenced in 2014 (MRC 2014). Understanding the opportunities and barriers faced by NGO actors through the Xayaburi dam consultation process can provide lessons for consultation processes of further hydropower dams on the Mekong River, and potentially for other rivers.

By comparing the advocacy strategies over the Mekong River of two NGO coalitions working under different rules and norms, in different countries, this book also highlights how differences in rules and norms can influence advocacy strategies. In particular, comparing NGO coalitions in Cambodia and Vietnam will bring a unique understanding of the influence of formal and informal rules and norms on civil society sectors both experiencing internal conflicts and adopting different political systems.

References

263 NGOs 2011. Global call to cancel the Xayaburi dam on the Mekong River mainstream in Northern Lao PDR. A letter addressed to Thongsing Thammavong, the Prime Minister of Lao PDR and Abhisit Vejjajiva, the Prime Minister of Thailand.

Almond, Gabriel A. Jr., G. Bingham Powell, Kaare Strøm, and Russel J. Dalton. 2004. *Comparative Politics Today: A World View*, 8th edn. London: Pearson Longman.

Betsill, Michele M. 2008. Environmental NGOs and the Kyoto Protocol negotiations: 1995 to 1997. In *NGO Diplomacy: The Influence of Nongovernmental Organizations in International Environmental Negotiations*, edited by M. M. Betsill and E. Corell. London: MIT Press.

Betsill, Michele M., and Elisabeth Corell. 2008. Introduction to NGO diplomacy. In *NGO Diplomacy: The Influence of Nongovernmental Organizations in International Environmental Negotiations*, edited by M. M. Betsill and E. Corell. London: MIT Press.

Bombay, Peter. 2001. The role of environmental NGOs in international environmental conferences and agreements: some important features. *European Environmental Law Review* 7: 228–231.

Borneo Project. 2012. *NGO coalition condemns Malaysian dam plans*. http://borneoproject. org/updates/ngo-coalition-condemns-malaysian-dam-plans accessed 24 February 2015.

Brown, L. D., S. Khagram, M. H. Moore, and P. Frumkin. 2000. *Globalization, NGOs and Multi-sectoral Relations*. Working Paper No. 1. Cambridge, MA: Hauser Center for Nonprofit Organizations and Kennedy School of Government, Harvard University.

Bruch, Carl. 2005. Evolution of public involvement in international watercourse management. In *Public Participation in the Governance of International Freshwater Resources,* edited by C. Bruch, L. Jansky, M. Nakayama and M. A. Salewicz. Tokyo: United Nations University.

Casey-Lefkowitz, Susan, William J. Futrell, Jay Austin, and Susan Bass. 2005. The evolving role of citizens in environmental enforcement. In *Making Law Work: Environmental Compliance & Sustainable Development, Volume 1*, edited by D. Zaelke, D. Kaniaru and E. Kružíková. London: Cameron May.

Charnovitz, Steve. 1997. Two centuries of participation: NGOs and international governance. *Michigan Journal of International Law* 18: 183–286.

—— 2006. Nongovernmental organizations and international law. *American Journal of International Law* 100 (2): 348–372.

Ebbesson, Jonas. 2007. Public participation. In *The Oxford Handbook of International Environmental Law*, edited by D. Bodansky, J. Brunnée and E. Hey. Oxford: Oxford University Press.

Edwards, Michael. 2004. *Civil Society*. Cambridge: Polity Press.

Finnemore, Martha, and Kathryn Sikkink. 1998. International norm dynamics and political change. *International Organization* 52: 887–917.

Fischer, William F., ed. 1995. *Toward Sustainable Development: Struggling Over India's Narmada River*. New York: M.E. Sharpe.

Florini, A., and P. J. Simmons. 2000. What the world needs now? In *The Third Force: The Rise of transnational civil society*, edited by A. Florini. Tokyo. Washington, DC: Japan Center for International Exchange, Carnegie Endowment for International Peace.

Fox, Leslie M., and Priya Helweg. 1997. *Advocacy Strategies for Civil Society: A Conceptual Framework and Practitioner's Guide: Final Draft*. Report prepared for USAID. www. innonet.org/resources/files/Advocacy_Strategies_for_Civil_Society.pdf accessed 18 April 2013.

Fu, Teng. 2009. *Dams and Transnational Advocacy: Political Opportunities in Transnational Collective Action*. Washington, DC: Catholic University of America.

Gemmill, Barbara, and Abimbola Bamidei-Izu. 2002. The role of NGOs and civil society in global environmental governance. In *Global Environmental Governance*, edited by D. C. Etsy and M. H. Ivanova. New Haven, CT: Yale School of Forestry & Environmental Studies.

Ghosh, Sujay. 2009. NGOs as political institutions. *Journal of Asian and African Studies* 44 (5): 475–495.

Ho, Peter. 2007. Embedded activism and political change in a semiauthoritarian context. *China Information* 21 (2): 187–209.

Hunter, David. 2005. The emergence of citizen enforcement in international organizations. In *Making Law Work: Environmental Compliance & Sustainable Development*, edited by D. Zaelke, D. Kaniaru, and E. Kružíková. London: Cameron May.

ICEM. 2010. *MRC Strategic Environmental Assessment (SEA) of the Hydropower on the Mekong mainstream*. Hanoi, Vietnam: Mekong River Commission. www.mrcmekong.org/ assets/Publications/Consultations/SEA-Hydropower/SEA-Main-Final-Report.pdf accessed 3 May 2014.

Keck, Margaret E., and Kathryn Sikkink. 1999. Transnational advocacy networks in international and regional politics. *International Social Science Journal* 51 (159): 89–101.

Koh, Harold Hongju. 1998. 1998 Frankel Lecture: Bringing international law home. *Houston Law Review* 35.

Millennium Ecosystem Assessment. 2005. *Overview of the Millennium Ecosystem Assessment*. www.millenniumassessment.org/en/About.html accessed 7 June 2015.

MRC. 2014. *Don Sahog Hydropower Project*. Mekong River Commission. www.mrcmekong. org/news-and-events/consultations/don-sahong-hydropower-project/ accessed 22 December 2014.

Ohanyan, Anna. 2012. Network institutionalism and NGO studies. *International Studies Perspectives* 13 (4): 366–389.

TRAFFIC. 2008. *About TRAFFIC*. www.traffic.org/overview/ accessed 13 April 2011.

UNFCCC. 2011. *Side Events and Exhibits. COP17/CMP7*. United Nations Framework Convention on Climate Change. https://unfccc.int/files/meetings/durban_nov_2011/ application/pdf/see_brochure_cop_17_cmp_7.pdf accessed 9 January 2014.

United Nations. 1992. *Agenda 21*. www.un.org/esa/sustdev/documents/agenda21/english/ Agenda21.pdf accessed 24 February 2010.

Part I
Research design

2 Analysing NGO advocacy strategies

Introduction

NGOs adopt a wide range of strategies in advocacy work. This chapter provides an overview and analysis of existing theoretical approaches used in the analyses of NGOs' advocacy strategies, including approaches examining the influence of rules and norms. This section leads to the identification of the methodology used for this book, which is then discussed in Chapter 3. As discussed earlier, this book focuses on understanding how rules and norms influence the advocacy strategies of NGO coalitions. The chapter therefore discusses existing theoretical approaches used to understand how rules and norms affect actors' behaviours, and concludes with a discussion of key issues identified in the literature that are used as a basis for developing the specific analytical approach of the book.

Understanding rules and norms

Rules and norms are the central concepts used in this book. These terms are used in different ways within different disciplines. For example, in legal disciplines the word 'norm' is used in a similar way to 'rule', and these terms are often used to define each other: *Black's Law Dictionary* defines rule as 'a general norm mandating or guiding conduct or action in a given type of institution' (Garner 2004: 1357), and the *Dictionary of Law* defines 'norm' as 'an authoritative standard or rule of behaviour' (Curzon 1998: 257). In sociology, the term 'norm' has less formal status than 'rule', and is defined as 'expectations about appropriate conduct which serve as common guidelines for social action' (Abercrombie *et al.* 2000: 243) or 'a shared expectation of behaviour that connotes what is considered culturally desirable and appropriate' (Scott and Marshall 2005: 451). The authors of the latter definition recognize the similarities of these terminologies, but distinguish between rules and norms. Norms are similar to rules or regulations in being prescriptive, although they lack the formal status of rules.

Ostrom provides a definition of rule as 'shared understandings by participants about enforced prescriptions concerning what actions (or outcomes) are required, prohibited, or permitted' (Ostrom 2005: 18). Ostrom refers to Black's

(1962) four classifications of how 'rule' is used in everyday conversations: regulations, instructions, precepts, and principles, and claims her definition of rule is based on this understanding of 'regulations' (Ostrom 2005: 16; Black 1962). Rules in relation to regulation refer to the fact that rules are 'announced, put into effect, enforced, disobeyed, broken, rescinded, changed, revoked, and reinstated' (Ostrom 2005: 19). Ostrom's concept of rules was developed based on her studies of communities and their use of resources, usually within well defined boundaries. Thus her study focuses on 'rules in use' defined as 'the set of rules to which participants would make reference if asked to explain and justify their actions to fellow participants' (Ostrom 2005: 19). Concepts such as 'culture' or 'value' that provide principles for guiding participant's actions are not integrated into her definition of rules, but rather included as 'attributes of the communities' (Ostrom 2005: 27) in her framework of analysis. Ostrom's framework is discussed in a later section of this chapter.

Williamson (2000), one of the founding scholars of new institutional economics, does not provide specific definitions of rules and norms. However, he provides definitions of different levels of institutions which include norms and rules. As the primary concern of institutional economists is in understanding humanly devised constraints that affect economic transactions (North 1991: 97–98), Williamson's (2000) categorization of an institution is based on the types of transaction in which it engages, as well as its frequency of change, as illustrated in Table 2.1. The first level is the informal institution, which is embedded in society, including customs, traditions, norms and religions. This level of institution is typically formulated over 100 to 1000 years (Williamson 2000: 596–597). The second level is institutional environment, which refers to the formal rules (constitutions, laws, property rights), as well as instruments for governance, including executive, legislative, and bureaucratic functions of government, and the distribution of power across government. Williamson claims that institutions at this level take 10 to 100 years as a possible cycle of formulation and change (Williamson 2000: 597–598). The third level is the institution supporting economic governance, such as contracts. The cycle for change at this level is one to ten years (Williamson 2000: 579–599). The final level relates to the actual economic transaction activities regulated through all levels of institution described above, including employment, price, quantities, and incentives, where changes occur on a continuous basis (Williamson 2000: 597).

Williamson argues that higher levels of institution impose constraints on the institutions immediately below them (Williamson 2000: 596). This would mean that norms, which are part of first-level institutions, constrain formal rules, which are part of the second level. Williamson does not provide a specific location where informal rules are found in his categorization. However, as rules are one of the key components of institutions identified by new institutional scholars, informal rules can be considered as part of informal institutions which are included in the first level of institutions.

Ostrom's and Williamson's definitions allow us to consider both norms and rules as principles that guide actors' behaviour. Norms apply to wider concepts

Table 2.1 Economics of institutions

Level of institution	Type of institution*	Frequency of change (years)	Relevant theory
1	Embeddedness: Informal institutions, customs, traditions, norms, religion	100–1000	Social theory
2	Institutional environment: Formal rules of the game, especially property (polity, judiciary, bureaucracy)	10–100	Economics of property rights, positive political theory
3	Governance: Play of the game, especially contract (aligning governance structures with transactions)	1–10	Transaction cost economics
4	Resource allocation and employment (prices and quantities, incentive alignment)	Continuous	Neoclassical economics, agency theory

Source: Adapted from Williamson (2000: 597).

Note
* Except level 4, which refers to economic transaction activities.

including culture and values, whereas rules tend to have more focused applicability to specific target groups. This book defines norms as the 'shared expectations and guiding conduct of behaviour that are considered appropriate within the given society'.

Rules are defined as principles regulating practice or procedure within a given society. Rules can be formal or informal (Lowndes 1996: 193). Lowndes argues that

> Formal rules are consciously designed and clearly specified – as in the case of written constitutions, contractual agreements, property rights, the terms of reference and standing orders of committees, and so on. Informal rules are not consciously designed or specified in writing – they are the routines, customs, traditions and conventions that are part of habitual action.
>
> (Lowndes 1996: 193)

The word formal refers in this respect to 'following established procedural rules, customs, or practices' (Garner 2004: 678) or 'suitable for or expected on formal occasion' (Allen 2000: 551). Formal rules include laws, regulations and policies as well as, in this case, rules formally adopted and codified by NGO coalition members. Law is defined as 'the body of rules, standards, and principles that the courts of a particular jurisdiction apply in deciding controversies brought before them' (Garner 2004: 900) or 'the written and unwritten body of rules largely derived from custom and formal enactment which are recognised as binding among those persons who constitute a community or state' (Curzon 1998: 215). In essence it is a type of rule that is often adopted through the legislative body of the country. The relationship between laws and norms is important to consider, as suggested by legal theorist Hans Kelsen in his 'pure theory of law', which discusses that laws are norms, and 'a society's legal system is made up of its norms, and each legal norm derives its validity from other legal norms' (Garner 2004: 1086). In other words, law is often created or adopted when a norm is adopted by a society as it reaches an internalization process, where the norm is accepted by society (Finnemore and Sikkink 1998: 895). The terms 'law' and 'regulation' are used similarly: law is a 'rule of conduct formally recognized as binding or enforced by authority' (Allen 2000: 792); regulation is 'a rule or order, having legal force, usually issued by administrative agency' (Garner 2004: 1311). The word policy in the context of public policy is used as the 'general principles by which a government is guided in its management of public affairs' (Garner 2004: 1196). This definition positions policy as a type of rule that is the 'principle regulating practice or procedure, a dominant custom or habit' (defined earlier), which specifically guides a government in its management of public affairs. Therefore policy can be considered as part of 'rule' in a broad context. This book uses the terminology of 'formal rule' to mean 'the rules which are established through recognized or official process'.

The term 'informal' is defined as 'not done or made according to a recognized form, irregular, unofficial, unconventional' (Brown 1993: 1364). Therefore informal rule in this book refers to rules established without official process that are applicable to the given society. In this book's case studies, informal rules include informal working arrangements or working culture among NGO coalition members, or informal rules which authorities impose in exercising their power. While the term 'norm' refers to shared expectation among wider society, informal rules and norms are discussed as a same category in this book as there are overlaps in these two concepts.

Understanding advocacy strategies

Dictionaries define advocacy as 'the function of an advocate' (Brown 1993: 32; Allen 2000: 19), defining advocate as 'somebody who pleads the cause of another before a tribunal or court' or 'somebody who defends or supports a cause or proposal' (Allen 2000: 19). These definitions indicate that advocacy is an act of promoting a certain position. In the literature targeted at NGOs and grassroots organizations, the definition of advocacy is often linked to policy change. An advocacy toolkit published by CARE, an international humanitarian NGO, defines advocacy as 'a deliberate process of influencing those who make policy decisions' (Sprechmann and Pelton 2001: 2). An advocacy guide for civil society commissioned by USAID defines advocacy as 'the process in which a group(s) applies a set of skills and techniques for the purpose of influencing public decision-making: the ultimate result is to achieve a well defined social, economic or political policy goal or reform' (Fox and Helweg 1997: 13). As this book examines advocacy strategies aimed at human development and the environment, advocacy in the context of this book is defined as a process or an act attempting to influence decision-making which affects the public at large.

Moving on to the concept of strategies, dictionaries define the term strategy as 'long-term planning in the pursuit of objectives, or the art of this' (Allen 2000: 1391), or 'the art of skill of careful planning towards an advantage or a desired end' (Brown 1993: 3085). In the literature, as well as in practical toolkits for practitioners developing advocacy strategies for NGOs and grassroots, it is usually suggested that advocacy strategies should contain: 1) goal of advocacy; 2) target audiences; 3) activities and tactics; 4) allies and opportunities; 5) success indicators, and 6) timeframe (Sprechmann and Pelton 2001; Gordon 2002; UNICEF 2010; Wells-Dang 2011). Referring to these key elements of advocacy strategy, this book includes advocacy goals, target audiences, activities, and tactics used by the NGO coalitions. The conceptual framework of the relationship between these elements is illustrated in Figure 2.1, which provides a way to categorize various strategies of NGO coalitions, discussed later in this book. Figure 2.1 is used for this purpose in Chapter 7, after providing an overview of various strategies undertaken by the case study NGO coalitions.

Understanding activities and tactics is a key aspect of understanding advocacy strategies. The following sections discuss key approaches in understanding these tactics, which have particular relevance to the strategies adopted by the case study coalitions. They include: 1) use of networks; 2) use of science; and 3) use of media.

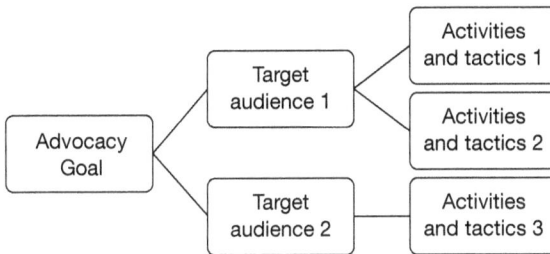

Figure 2.1 Framework for understanding advocacy strategies

Source: Developed by the author from Sprechmann and Pelton (2001); Gordon (2002); UNICEF (2010); Wells-Dang (2011).

Strategy 1: Use of networks

In the simplest form, a network is 'a collection of points joined together in pairs by lines' (Newman 2010: 1). The word 'network' is used in different disciplines including mathematics, computer science, biology, and social science (Newman 2010: 1–10). In social science, a network often refers to human beings or social entities, and is referred to as a social network. A social network is 'a set of relations that apply to a set of actors, as well as any additional information on those actors and relations' (Prell 2012: 9). The analysis of social networks involves understanding relationships amongst actors (Wasserman and Faust 1994: 3).

Referring to social networks associated with policy processes, political scientists often use the term 'policy network', which is referred to as 'sets of formal and informal institutional linkages between governmental and other actors structured around shared interests in public policy-making and implementation' (Rhodes 2007: 1244). Rhodes (1997) provides different typologies of networks involved in the policy process, depending on the characteristics of the network. For example, a policy community is characterized by the stability of the relationships and a highly restricted membership, whereas an issue network has a large number of participants with a limited degree of interdependence, and is often seen as a more unstable network (Rhodes 1997: 38, 44).

The word 'coalition' is used in a similar way to 'network'. Yanacopulos discusses the difference between a network and a coalition, indicating that coalition refers to more permanent links than those forged by a single-issue thematic network (Yanacopulos 2005: 95). According to Yanacopulos, coalitions often have permanent staff, a more permanent membership base, and a secretariat. More importantly, coalitions often share broader strategic goals than a single issue-based network (Yanacopulos 2005: 95). Referring to networks, coalitions, and movements in a transnational context, Fox differentiates between these based on different degrees of shared characteristics, as described in Table 2.2 (Fox 2010). Referring to these characteristics, Fox (2010) claims that coalitions need shared targets and require clear terms of engagement to be sustainable. Sabatier and Jenkins-Smith (1999) refer to an advocacy coalition as something consisting of actors from a variety of organizations who share a set of policy beliefs (Sabatier and Jenkins-Smith 1999; Weible *et al.* 2009: 122). An advocacy coalition is 'composed of people

Table 2.2 Transnational civil society networks, coalitions, and movements

Shared characteristics	Transnational networks	Transnational coalitions	Transnational movement organizations
Exchange of information and experiences	Shared	Shared	Shared
Organized social base	Some have bases, others do not	Some have bases, others do not	Counterparts have bases
Mutual support	Sometimes shared, from afar, sometimes strictly discursive	Shared	Shared
Material interests	Not necessarily shared	Sometimes shared	Sometimes shared
Joint actions and campaigns	Sometimes loose coordination	Shared, based on mutually agreed goals, often short-term, tactical	Shared, based on shared long-term strategy
Ideologies	Not necessarily shared	Not necessarily shared	Usually shared

Source: Adapted from Fox (2010: 488).

from various governmental and private organizations that both 1) share a set of normative and causal belief and 2) engage in a nontrivial degree of coordinated activity over time' (Sabatier and Jenkins-Smith 1999: 120). In summary, a coalition is a type of network comprised of actors who have shared objectives and conduct joint actions.

Networks as a strategic approach adopted by NGOs

The use of networks is often a key approach adopted by NGOs and civil society actors in their advocacy work (UNICEF 2010: 95). Strategic use of networks allows NGOs and civil society actors to gain the ability to influence decision-making and policy. Lenschow utilized a policy network framework to examine the role of NGOs in environmental policy, arguing that while environmental NGOs are often regarded as peripheral actors, they can play an influential role in policy-making through their strategic network-building (Lenschow 1997). Haas (1992) claims that NGOs promote norms, and provide knowledge through taking part in an epistemic community (Haas 1992: 3; 2007: 793). An epistemic community is a network of experts who have authoritative claims within their domain of expertise (Haas 1992: 3; 2007: 793). Epistemic communities play important roles in the norm internalization process (Koh 1998: 646), which is the stage when the norm acquires the quality of being taken for granted by society (Finnemore and Sikkink 1998: 895), illustrating the importance of networks. Being part of an epistemic community allows NGOs to influence public decision-making as experts. Haas (2007), however, claims that comparatively few NGOs are drawn to epistemic communities, since NGOs tend to value principled belief over causal beliefs in policy discussions, while an epistemic community maintains its expert authority through a seemingly neutral scientific understanding of a particular subject (Haas 2007: 793).

As suggested by Rhodes (2007), networks are often created when organizations are dependent on each other for resources, and where they need to exchange resources in order to achieve their goals (Rhodes 2007; Parsons 1995: 186). Resources, in this context, include human resources, financial resources, and physical assets, as well as knowledge. The network approach became attractive to political scientists in the 1970s, allowing them to analyse how people interact within a policy arena, using a flexible approach, as policy is often a product of the interaction of many actors (Parsons 1995: 185). Network theorists take the view that policies emerge as a result of bargaining or cooperation between different network members (Rhodes 1997; Hill 2009). The network approach focuses on understanding the patterns of formal and informal relationships among network members engaged in a policy process, including both governmental and nongovernmental actors (Parsons 1995: 185). It is an advancement of the pluralist and corporatist theory that claims power is distributed widely throughout society (Hill 2009; Rhodes 1997: 37). The network approach is used particularly in explaining the shift from 'government' to 'governance' and the engagement of nongovernmental entities in the form of governance (Rhodes 1997). Networks of NGOs, therefore, can potentially contribute towards improved governance.

In the contemporary world, where policy-making is no longer the sole responsibility of government actors, but one in which diverse nongovernmental actors are involved, the conceptualization of governance places emphasis on interdependence between governmental and nongovernmental bodies (Lovan *et al.* 2004: 3). Rhodes defines one aspect of governance as 'self-organizing, inter-organizational networks', arguing that governance takes place through exchange of resources among actors that include both governmental and private actors (Rhodes 1996: 658-660). Rhodes continues to discuss governance as inter-dependence between organizations, and a process of self-organization with limited accountability to the state (Rhodes 1996: 660). Thus networks are one of the key contributors to the development of governance, and NGOs are one of the important players in this process.

Understanding networks can shed light on understanding society. While social scientists typically study the networks of human beings in order to understand society, scholars of actor network theory (ANT) take the approach that society is composed of networks of both human and nonhuman entities. Therefore it is vital to understand the interactions of both in order to understand society (Callon 2001: 62; Dolwick 2009). Referring to the literature on civil society, Wells-Dang suggests that civil society as a whole can be understood as a network (Wells-Dang 2011). As networks are one of the key forms that NGOs and civil society actors adopt when conducting advocacy work, understanding factors affecting the way networks operate becomes important in order to understand how wider participation and governance over environmental and natural resources management might be developed.

Transnational advocacy networks

NGOs and activists advocating changes and improvements in a variety of international issues often establish a network, which is described by Keck and Sikkink (1998) as a transnational advocacy network (TAN). A TAN is a transnational network of NGOs and other actors committed to championing the causes of others or defending a common cause or proposition (Keck and Sikkink 1998: 8). The network may include: 1) international and domestic NGOs, research and advocacy organizations; 2) local social movements; 3) foundations; 4) the media; 5) churches, trade unions, consumer organizations, and intellectuals; and 6) parts of regional and international organizations (Keck and Sikkink 1998: 9). Due to the diversity of actors engaged, a TAN is often capable of identifying a variety of venues through which to raise the issues of concern, utilizing the strength of various actors. As Keck and Sikkink indicate:

> Transnational value-based advocacy networks are particularly useful where one state is relatively immune to direct local pressure and linked activists

elsewhere have better access to their own governments or to international organizations. Linking local activists with media and activists abroad can then create a characteristic 'boomerang' effect, which curves around local state indifference and repression to put foreign pressure on local policy elites. Activists may 'shop' the entire global scene for the best venues to present their issues, and seek points of leverage at which to apply pressure. Thus international contacts amplify voices to which domestic governments are deaf, while the local work of target country activists legitimizes efforts of activists abroad.

(Keck and Sikkink 1998: 200)

Keck and Sikkink (1999) suggest different types of strategies that TANs may adopt in their advocacy work; one of these strategies is termed information politics, which is the ability of the network to gain and mobilize information that garners political attention (Keck and Sikkink 1999: 95). The second strategy is symbolic politics, which is the ability to frame a situation symbolically in a way that resonates with audiences that are physically removed from the situation of the subject of advocacy (Keck and Sikkink 1999: 96). The third type of strategy is termed leverage politics, the ability of a network to influence more powerful institutions to which weak groups would not have otherwise gained access (Keck and Sikkink 1999: 97). The final type of strategy suggested is accountability politics, a strategy that holds governments responsible for commitments which they made previously (Keck and Sikkink 1999: 97).

One of the key issues that networks face is the relationship and interaction among the network members. For a TAN, the issue of managing relationships among its network members is particularly challenging as it typically includes members from different countries with diverse backgrounds. A TAN often includes organizations from the North working with communities where advocacy issues are taking place in the South. Members from the North tend to have more access to resources, including financial and technical resources, creating an uneven landscape (Meierotto 2009: S283). On the other hand, members from the North often gain legitimacy for their actions through their contacts and support to civil society groups in the South (Keck and Sikkink 1999; Hudson 2001: 331). Yanacopulos points out legitimacy as one of the important resources of nonprofit organizations (Yanacopulos 2005: 97). Conducting an advocacy campaign in a democratic manner within networks is also important, particularly as TANs tend to involve people and organizations in unequal positions (Jordan and van Tuijl 2000: 2051). Commitment to this democratic principle is referred to as political responsibility by Jordan and van Tuijl.

Referring to different types of political responsibilities that exist among existing transnational NGO campaigns, Jordan and van Tuijl (2000) provide four types of relationship: 1) cooperative campaigns where members share interlocking objectives and the network achieves an optimal level of political responsibility towards the most vulnerable members; 2) concurrent campaigns where members have different but compatible objectives without achieving a high level of political responsibility; 3) disassociated campaigns where NGOs engaged have parallel representation and conflicting objectives, where the campaign achieves a low level of political responsibility; and 4) competitive campaigns where the advocacy of certain actors may have adverse or counterproductive impacts on other actors (Jordan and van Tuijl 2000: S2056–S2061). These types of relationships among advocacy campaign actors have informed the analysis of relationships between various NGOs/civil society coalitions.

Similarly, referring to the ways in which NGOs attempt to strengthen community-based and grassroots institutions, Covey provides three typologies of such alliances. One is a grassroots-centred alliance where policy questions are defined from a grassroots point of view; while the second type of alliance is the NGO-centred alliance where NGOs define agendas and strategies of alliances (Covey 1995). The third type of alliance is a mixed alliance where a variety of actors formulate the alliance; these actors may include the poor, the middle class and elites, and may face difficulties in bringing benefits to all groups (Covey 1995). While, in reality, it is not easy to identify clearly where the original agenda and advocacy strategy is created, these typologies also provide a lens for analysing the working relationships among network members.

In conclusion, the literature provides three key lessons that can be used in analysing networks. First, understanding the resource dependency of networks can shed light on the type and nature of the network. Second, understanding the characteristics of network members, their resources, relationships, and interactions, is important for understanding the strategy development of the network, and thus is an important aspect to consider when analysing the strategies of NGO coalitions. Finally, analysis of the literature identified networks as one of the key forms of governance, and NGOs as one of the most important actors.

Strategy 2: Use of science

Linking science into policy- and decision-making is recognized as an important and undisputable aspect of public policy-making (Pielke and Klein 2010: 14, 19; Huitema and Turnhout 2009: 576). Some NGOs have the expertise to provide policy-relevant scientific knowledge to decision-makers (Gemmill and Bamidei-Izu 2002: 11). However, the science–policy interface is not easy as it requires an attempt to bring two different worlds together. The world of science, on one hand, is considered by many to be objective, neutral, independent, factual, and based on standardized methodology. The world of policy-making, on the other hand, is considered by some to be subjective, based on ideology and values, and one that takes an opportunistic approach (Huitema and Turnhout 2009: 579). Scientific research often requires long-term investment to gain results, and often results in uncertainties, whereas policies and decisions have shorter life-cycles, are often dependent on election results, and require decision-making even when scientific uncertainties exist (De Santo 2010: 414, 418; Van der Sluijs 2005: 87). For scientific knowledge to be used for policy-making, it needs to be translated and transferred to usable knowledge within the policy domain (Turnhout *et al.* 2007: 220). Because of the divergence between science and policy, the role played by actors who translate and transfer scientific knowledge to the policy world becomes important. Organizational actors who sit at the boundary of these different worlds are called 'boundary organizations' (Guston 2001: 400). Boundary organizations exist at the frontier of politics and science, and facilitate participation and collaboration between scientists and non-scientists (Guston 2001: 401).

The questions 'what is science' and 'what distinguishes science and non-science' are important when considering the science–policy interface. Academic literature refers to this demarcation of science and non-science as 'boundary work' (Gieryn 1983: 399; Guston 2001; Turnhout *et al.* 2007). The concept of 'boundary work' was originally proposed by Gieryn (1983), who reviewed how science was perceived in historical contexts and concluded that 'boundaries of science (and non-science) are ambiguous, flexible, historically changing, contextually variable, internally inconsistent and sometimes disputed' (Gieryn 1983: 792). One example that illustrates Gieryn's conclusion is the disciplinary battles between anatomists and phrenologists in early nineteenth-century Edinburgh, where established scientists (in this case, anatomists) refused to accept new science (phrenology) which was threatening the reputations and authorities already established by anatomists, and which in turn created various barriers for phrenologists in conducting their academic work (Gieryn 1983: 789). Illustrated by this example, boundaries of science and non-science can be seen as socially constructed (Turnhout *et al.* 2007: 221). The ambiguity and flexibility of boundary work may lead to confusion as well as to a risk of the politicization of science (Guston 2001).

The literature on the use of science brings several key aspects to attention in analysing NGOs' use of science as advocacy strategy. First, when considering NGOs' role as science providers, NGOs' relationships with boundary organizations become an important factor. Second, the mandate and ability of the boundary organization to communicate science is also important. Third, the literature indicates that what distinguishes science and non-science could be subjective and dependent on who evaluates scientific information. Again, the role of boundary organizations, and the credibility of science by NGOs in the eyes of boundary organizations and government actors, are critical when analysing the use of science in NGOs' strategies.

Strategy 3: Use of media

The mass media play various roles within modern society. Most people rely on the mass media as the source of information about events and political affairs occurring both locally and globally (Negrine 1994: 1). The images, texts, and their interpretation provided by the media form the basis for public perception of events (Graber 2002: 6-11; Negrine 1994: 2). The media also plays an important role in socialization, a process where individuals internalize values, beliefs, and culture in a particular society (Croteau and Hoynes 2003: 13; Graber 2002: 2). According to Meyer, through balanced, objective, comprehensive reporting, the mass media have a capacity to contribute to democratic communication (Meyer 2002: 4). Graber also claims that media stories are often indications of what is considered important in society (Graber 2002: 2).

The mass media also have the capacity to shape public policy, and at times to manipulate politics (Spitzer 1993: 5; Graber 2002: 12). When the media report on news and events, they frame the stories, providing meanings to events (Scheufele 1999: 106–107). While the professional ethics of journalism may oblige journalists to cover stories objectively, they also exercise discretion over how they frame the stories (Simon and Jerit 2007: 258). Through framing, the mass media have the capacity to build up public images of, for example, political figures, and can force the public to pay attention to certain issues (McCombs and Shaw 1972: 177). As discussed earlier, framing is also an important way for certain norms to be accepted by society (Finnemore and Sikkink 1998: 897).

In a reverse way, politics and political systems are also capable of manipulating the media. Politicians may take advantage of journalists' discretion in framing certain issues, and may encourage them to frame stories in a way that suits their political agenda (Simon and Jerit 2007: 258). The type of political regime can also affect the way in which the media produce and communicate stories. Democratic regimes tend to emphasize the importance of press freedom (Negrine 1994: 25; McQuail 1987: 113), and the media in a democratic regime are expected to work on the assumption that the press is an instrument used to check and scrutinize the actions of the government. Thus the press should provide critical thinking on major policies, sometimes challenging government policies when they seem flawed (Graber 2002: 20; Negrine 1994: 25). The selection of news and entertainment programmes, however, is usually based on audience appeal (Negrine 1994: 35; Graber 2002: 20). In contrast, authoritarian and socialist regimes assume that the government knows and respects people's best interests, therefore the media should not attack the government and its policies, but rather should play a supportive role (Graber 2002: 20; McQuail 1987: 111). It also contends that the selection of news and entertainment programmes should be based on the social values of the regime (Graber 2002: 20). What is to be broadcast and what is not broadcast is often determined by leading political parties (Young 2012: 3–4).

Governments can also control the media through regulations. Governments may require a permit to start a business in the media, or may place restrictions on ownership (Graber 2002: 26). Media with a limited broadcast spectrum, such as radio and TV, need allocation of the spectrum, which is often controlled by the government or semi-governmental bodies (Graber 2002: 26; Croteau and Hoynes 2003: 78). In a totalitarian regime, governments own and operate the media, not allowing other media to exist in the country, and enforce this by jamming channels and prohibiting the import of printed materials (Croteau and Hoynes 2003: 77; Graber 2002: 27). And in democratic regimes, where freedom of expression is guaranteed, certain types of regulations exist to ensure the media do not infringe their social responsibility, including regulation of broadcast content (Negrine 1994: 34; Croteau and Hoynes 2003: 101–107; Graber 2002: 19).

Using the media is one of the key advocacy strategies often adopted by NGOs and civil society actors, in order to raise public awareness (Sprechmann and Pelton 2001: 90). As discussed earlier, the mass media offer an effective way to shape public opinion on certain issues, and could potentially influence politics (Spitzer 1993: 5; Graber 2002: 12). Through these functions, the media provide pathways for advocacy actors to disseminate their messages and seek wider support for issues that are the subject of their advocacy work. Using the media allows advocacy organizations to deliver their message to a large number of people and potentially attract public support (Anderson 1997; Sprechmann and Pelton 2001: 90). Gaining media attention can also potentially increase credibility with policy-makers (Bunn and Ayer 2004: 3; Sprechmann and Pelton 2001: 90).

There are also challenges associated with the use of media in NGOs' advocacy efforts. Media coverage of advocacy events works as a validation of their importance (Gamson 2004: 252). However, coverage by the media does not necessarily provide an opportunity for advocacy groups to voice their opinions (Gamson 2004: 251). Depending on how stories are framed, there is also a risk that media attention may negatively affect the advocacy work or the organization working on these issues (Sprechmann and Pelton 2001: 90). It is also difficult for civil society advocacy groups to gain the status of regular 'standing' for journalists, which means the main sources of information for journalists. Journalists often tend to refer to authorities for their regular sources of information (Gamson 2004: 251). In regimes where the media are controlled by government, journalists may be reluctant, or not allowed, to report on certain issues, which hinders advocacy using the media.

In conclusion, using the media can be an effective tool for civil society coalitions' advocacy strategies in raising public awareness on certain issues. However, there are various opportunities and barriers associated with use of this strategy, depending on the political, social, and economic context (Graber 2002; McQuail 1987; Young 2012; Negrine 1994; Randall 1993). In analysing advocacy strategies using the media, it is important to note that, while the media have the potential to influence government and policies, government and its policies can also influence the media. Such influence can potentially affect NGO coalitions that choose to use the media within their advocacy strategy. The literature also discusses how political systems and regimes can affect the ways in which media and government interact.

Understanding NGOs

Nongovernmental organization (NGO) is a 'container concept' that describes what it is not (*non*governmental), rather than what it is, therefore the term

NGO is used and interpreted in many different ways (Büsgen 2006: 7; Fisher 1998: 5). *Black's Law Dictionary* defines an NGO as 'any scientific, professional, business or public-interest organization that is neither affiliated with nor under the direction of a government; an international organization that is composed of private individuals or organizations' (Garner 2004: 1080). The *Oxford Dictionary of Law* defines an NGO as 'a private international organization that acts as a mechanism for cooperation among private national groups in both municipal and international affairs, particularly in economic, social, cultural, humanitarian, and technical fields' (Martin and Law 2006: 375). Charnovitz (1997) claims that definitions of an NGO are applied more in an international context than in a domestic context. This is reflective of the origin of the term 'nongovernmental organization', which first appeared in the Charter of the United Nations (Charnovitz 1997: 186). These definitions of an NGO also broadly encompass most organizations that are not affiliated with government.

The term NGO, however, can also be defined by the purpose of the organization. Charnovitz provides a definition of NGOs as 'groups of individuals organized for the myriad of reasons that engage human imagination and aspiration, which can be set up to advocate a particular cause, such as human rights, or to carry out programmes on the ground, such as disaster relief' (Charnovitz 1997: 185–186). The terminology has a specific connotation when used in certain contexts: Fisher suggests that, in the developing nations, it generally refers to organizations engaged in development (Fisher 1998: 5). The term 'charitable organization' is used in a similar context, referring to organizations whose purposes are only for charitable causes. For instance, under Scottish law, charitable purposes may include poverty prevention, the advancement of education, religion, health, citizenship, community development, arts, heritage, culture, science, and public participation in sport (Article 7 *Charities and Trustee Investment (Scotland) Act* 2005). Although profit-making is typically not the primary objective of NGOs, the definitions discussed above do not exclude organizations that make profits as a result of their activities.

In this book, NGOs are defined as nongovernmental and non-profit organizations that are formulated to advocate certain particular aspirations for human development, environment, and many other charitable objectives. The book aims to understand the advocacy strategies of NGO coalitions, and to understand the rules and norms that affect their strategies. The research is conducted within the context of developing nations, where NGOs are mostly associated with aspirations for development and charitable objectives, rather than functioning as profit-making organizations. Here, 'NGO coalitions' are defined as coalitions of NGOs, or coalitions of NGOs and civil society actors coordinated by NGOs.

NGOs are usually seen as part of civil society; the term 'civil society' is used to refer to a wider sector within society which includes NGOs. The term is used to refer to aspects or groups of society that are distinct from states and markets (Steinberg and Powell 2006: 4). Therefore civil society is also referred to as the 'third sector', containing all associations and networks between family and state, except firms (Steinberg and Powell 2006: 4; Edwards 2004: vii). Cohen and Arato define civil society as 'a sphere of social interaction between economy and state, composed above all of the intimate sphere (especially the family), the sphere of associations (especially voluntary associations), social movements, and forms of public communication' (Cohen and Arato 1994: ix). Civil society, in the context of this book, utilizes these commonly adopted definitions and refers to the parts of society that are distinct from states and markets.

There are different ways to categorize NGOs, depending on the geographical scope of the NGO, the focus of the NGO's work, and the type of activity of the NGO. International NGOs work primarily at the international scale; national NGOs work primarily at the national scale; and community-based organizations (CBOs) consist of local community members and operate primarily within local communities (World Association of Non-Governmental Organizations 2010). International NGOs often base their headquarters in one country and have offices or branches in other countries where they operate. National NGOs and CBOs are often registered with national authorities, depending on applicable rules regulating the nongovernmental sector within each country.

NGOs can also be categorized by their types of activity. One type is advocacy NGOs, whose primary approach is in promoting specific causes through attempting to influence decision-making processes (World Bank 1998). Another type is development or operational NGOs, whose primary approach is in implementing projects or interventions in order to achieve development or charitable objectives. At times, the same NGO could take both types of approach, being an advocacy NGO at the same time as being a development NGO (World Bank 1998).

While NGOs are 'nongovernmental organizations', some NGOs are closely linked with government. Government-organized NGOs (GONGOs) are NGOs that are operated or established by governments, and have the nature of something between a government agency and an NGO (Wu 2003). This concept of GONGO contradicts the original definition of NGOs as being nongovernmental. They are at times created to achieve the objectives of a particular regime (Naim 2007). In the United Kingdom, quasi-autonomous NGOs (QUANGOs) are referred to as organizations that are funded by government, primarily functioning to deliver public services (Hogwood 1995; Almond *et al.* 2004: 197).

These different types and functions of NGOs are important to note when attempting to understand the characteristics of NGOs, which are the subject of this study. In particular, these distinctions are important in understanding how rules and norms could affect different types of NGOs.

Factors influencing NGOs

This section examines the existing literature that discusses the factors influencing NGOs, with particular focus on how rules and norms influence them, and how rules and norms might also influence advocacy strategies. As NGOs are part of civil society, the literature discussing the factors influencing both NGOs and civil society is included.

Some studies have examined the role of political regimes in influencing the activities of NGOs and civil society actors, claiming that the type of national political regime influences the nature of the state and its relationships with civil society. Through a study of enabling factors associated with state–civil society networks, Brinkerhoff claims that democratic regimes tend to progress further in terms of state–civil society networks, compared with authoritarian or limited democratic forms of government (Brinkerhoff 1999: 135). In his study, he refers to regime as a politico-bureaucratic setting (Brinkerhoff 1999: 135). Brinkerhoff goes on to discuss that a legal framework for the public sector that enables collaboration with civil society actors is also an important factor (Brinkerhoff 1999: 138). Similarly, Bryant (2001) studies how different regimes affect state–NGO relationships through the concept of critical engagement, which refers to a state–NGO relationship where both cooperation and conflict exist (Bryant 2001: 17). Comparing state–NGO relationships between liberal democratic regimes (through the example of the post-1985 Philippines) and authoritarian regimes (using the example of the late 1960s to late 1990s in Indonesia), Bryant claims that in both types of regime, critical engagement is used by both state and NGO actors as a way to enhance political control and legitimacy (Bryant 2001). Both studies indicate that a political regime, which is the system or rule of government (Brown 1993: 2527), shaped by 'implicit or explicit principles, norms, rules and decision-making procedures (Haggard and Simmons 1987: 493), affects the way NGO actors take their actions *vis-à-vis* state actors. The studies support the argument that rules and norms prevailing within a certain political context influence NGOs; however, they do not provide a detailed discussion of how rules, norms, and actors within a particular regime interact in creating the influence. Both studies also focus on understanding the influence of political regimes, and do not extend their analysis to the influence of the rules and norms that shape NGOs' behaviour within their sphere of activity.

Embeddedness is another concept that scholars have used in identifying the relationship between regimes and the behaviour of NGOs. The concept of 'embeddedness' has been developed through the work of scholars who attempted to explain economic phenomena from a sociological perspective, claiming that economic behaviour is embedded in society, and that understanding economic behaviour requires knowledge of history, culture, social relations, and networks (Granovetter 1985: 482). Williamson, one of the founding scholars of new institutional economics, refers to embeddedness as 'informal institutions', which includes norms, customs, traditions and religions (Williamson 2000: 596).

Scholars studying Chinese NGOs and the activism of their networks argue that NGOs operating in the Chinese context, where there is an authoritarian regime, take an 'embedded' approach (Gu 2008; Ho and Edmonds 2008;

Wells-Dang 2011). Through analysis of the inter-organizational collaboration of NGOs, Lawrence and colleagues suggest that embeddedness in an institutional environment is one of two important factors, along with the level of involvement among participants (Lawrence *et al.* 2002: 289). In this study, Lawrence refers to institutions as 'relatively widely diffused practices, technologies, or rules that have become entrenched in the sense that it is costly to choose other practices, technologies or rules' (Lawrence *et al.* 2002: 282). The study attempts to understand how NGOs' collaboration with other organizations can enhance the creation of 'proto-institutions' that are new practices, technologies, and rules (Lawrence *et al.* 2002: 283). The study indicates that although organizational collaboration is primarily a matter of interest among collaborating parties, if they do not take into account how the collaboration embeds them in wider institutional context, there is a risk of foregoing opportunities for creating proto-institutions (Lawrence *et al.* 2002: 289). This analysis provides useful lessons for NGOs' advocacy work as it implies that if NGOs intend to advocate for new ways of doing things, or to adopt new perspectives, it is important to take into consideration that their strategies need to be embedded in their society's institutional environment.

The studies that identify embeddedness as one of the factors affecting NGOs' behaviours suggest the importance of understanding the influence of norms on NGOs. However, further identifying the details of specific aspects of embeddedness, and how embedded actions occur in relation to formal rules, could help NGOs and civil society actors in the practical application of explaining and strategizing their future actions.

Resource dependency is another area that various scholars have studied as a factor affecting NGOs and civil society actors. Saxon-Harrold, studying British voluntary organizations, identified competition and resource dependence as factors that determine voluntary organization strategies (Saxon-Harrold 1990: 126). Hsu, studying state–NGO relationships in the Chinese context, explains this relationship as a result of resource dependency between the Chinese state and NGOs, where NGOs are in constant need of sustaining their finances, and view state agencies as resource-rich conglomerations (Hsu 2010). NGOs may start their activities as initiatives aiming for charitable objectives; however, once set up as organizations, like any other organization, NGOs need to secure a constant supply of resources to survive (Hsu 2010: 267). Referring to transnational campaigns, Yanacopulos (2005) reminds us that advocacy campaigns at the global level tend to be extremely expensive, and that no one organization can finance an entire campaign. This creates incentives for NGOs to take part in transnational advocacy networks.

Necessary resources for NGOs are not limited to funding, but also include human resources, legitimacy, and information (Yanacopulos 2005: 97). Legitimacy is defined by Lister (2003) and Suchman (1995) as 'generalized perception or assumption that the actions of an entity are desirable, proper, or appropriate within some socially constructed system of norms, values, beliefs, and definitions' (Suchman 1995: 574). Although NGOs are often considered to represent the public, and may play a role as 'bottom-up brokers', NGOs may in fact represent certain opinions and interests not necessarily reflecting public opinion at large (Bodansky 1999: 619). While NGOs are often valued for their connections with local communities and grassroots opinion, Meierotto claims that there are times when such connections are not grounded in reality (Meierotto 2009: S284). In the international policy-making arena, the question as to what would constitute a meaningful level of civil society participation still needs clarification (Gemmill and Bamidei-Izu 2002: 15).

Issues of legitimacy and accountability are important when NGOs and civil society actors form international networks for the purpose of advocacy. Jordan and van Tuijl use the concept of political responsibility in order to distinguish different types of TANs based on their relationships among actors in the network (Jordan and van Tuijl 2000: 2053). Political responsibility is a commitment to conduct an advocacy campaign with democratic principles in the process (Jordan and van Tuijl 2000: 2053). Similarly, Covey provides different typologies of NGO alliances from the perspective of the accountability and effectiveness of an NGO policy alliance (Covey 1995).

The legitimacy of NGOs is not only relevant to the public or the local populations which they attempt to represent, but is also important in the eyes of other actors. Lister examines the legitimacy of northern NGOs (NGOs from developed nations) and argues that these NGOs need to consider legitimacy in the eyes of a wide range of actors including donors, public, targets of advocacy, partner NGOs, southern governments, beneficiaries, and employees (Lister 2003: 186). International NGOs that encompass multiple entities and branch offices are often required to act within the remit of their organizational strategic priorities. These are shaped by various factors including the status of the subject matter, funding, public opinion, and relationships with their stakeholders.

Bodansky provides three main sources of legitimacy. First is the origin of the source, for example, historically a king or church has been seen as legitimate; second, legitimacy is associated with procedure – it is associated with the type of process that was followed by the people; and third, a successful outcome of an action can bring legitimacy to the public authority (Bodansky 1999: 612). While Bodansky discusses these sources of legitimacy in connection with legitimacy of authorities, the same perspective can apply for the legitimacy of NGOs.

These studies approach resource dependency from an organizational perspective, as based on the fundamental understanding that organizational behaviour is bounded in its context, and the survival of an organization depends on acquiring and maintaining resources (Pfeffer and Salancik 2003: 1–2). Resource dependency, however, can also be applied to the study of networks, as networks are often created to supplement resources among their members (Rhodes 1997). This perspective will be used to identify the incentives and motivations for the existence and operation of networks such as those studied in this book.

Some scholars have examined how institutional environments affect NGOs. Doh and Guay studied how institutional environments influence NGO activism, corporate social responsibility, and public policy in Europe and the USA, using a new institutionalism perspective and stakeholder analysis as research methods. The authors do not provide a specific definition of institutional environment in this article; however, they take a new institutionalist approach and consider key concepts of institution to be laws, policies, formal agreements, and behavioural norms (Doh and Guay 2006: 49). The authors state that

> As NGOs become more successful in expanding their membership and developing sophisticated networks to effect change, they become more integrated into the institutional environment in which they operate, and the more they will influence the formal institutional setting, contributing to a dynamic, co-evolutionary process.
>
> (Doh and Guay 2006: 54)

Building on the concept of institutional isomorphism, some scholars claim that institutions that interact with each other tend to become similar to each other (DiMaggio and Powell 1983; Ramanath 2009). Ramanath examines institutional isomorphism in the case of NGOs and claims that 'as more NGOs cooperate with the state, they become isomorphic in their structures and processes' (Ramanath 2009: 51). These studies indicate that NGOs change and adopt their operations and strategies depending on the institutional environment in which they work and the other organizations with which they collaborate, thus illustrating the influence of an institutional environment.

Recognizing the fact that NGOs take part in a network, Ohanyan (2012) proposes network institutionalism as a way to analyse NGOs in the context of international relations. In attempting to understand NGOs' behaviours in the context within which NGOs operate, Ohanyan claims that network institutionalism

> (i) transcends the state as a point of reference for NGO studies, (ii) examines NGO embeddedness into their immediate institutional environments and networks and (iii) acknowledges the diffusion of power within and between networks and treats power as a relational and dynamic attribute rather than a fixed construct.
>
> (Ohanyan 2012: 367)

Network institutionalism brings structure–agency perspective (discussed in the next section) into the relationship between networks and NGOs, and suggests that their relationships are mutually constitutive, and the relationship could be both cooperative and conflictual (Ohanyan 2012: 378). Network institutionalism is not an entirely new approach as it combines sociological institutionalism and network theory (Ohanyan 2012: 366), but it brings a useful approach to our understanding of individual NGOs' roles within an NGO coalition. In particular, the approach introduces perspectives that distinguish how NGOs influence structure (network, in this case) and *vice versa*, both positively and negatively (Ohanyan 2012: 381). The study does not provide a definition of institution. However, sociological institutionalism, on which network institutionalism is based, takes a wider approach in defining an institution as something that frames meaning and guides human actions, and includes 'formal rules, procedures or norms, symbol systems, cognitive scripts and moral templates' in its definition (Hall and Taylor 1996: 947).

In understanding factors affecting civil society actors, scholars of social movements may use the concept of political opportunity structures as factors that affect social movements (Tarrow 1998: 7; Sikkink 2005). These factors include

> 1) the nature of existing political cleavages in society; 2) the formal institutional structure of the state; 3) the informal strategies of the political elites *vis-à-vis* their challengers; and 4) the power relations within the party system (alliance structures).
>
> (Van der Heijden 1997: 27)

Political opportunity structures exist not only at the domestic level but also at the international level, which is relevant for social movements involving international issues and actors (Sikkink 2005: 156). Concepts of political structure provide useful insights for the analysis of institutional influence on advocacy strategies of NGO coalitions, in several ways. First, they provide both formal and informal aspects of political settings within which advocacy coalitions operate. Second, social movements typically aim to advocate on certain issues, which should also be taken into consideration when examining what may affect advocacy strategies. However, a scope for analysis focused on the political system and political actors (political context) does not integrate the influence of rules and norms adopted among members of social movement networks.

Influence of rules and norms

The review of existing literature identified several important lessons for analysing the influence of rules and norms on advocacy strategies of NGO coalitions. First, the existing literature argues that both rules and norms influence the behaviours of NGO actors. This is an important point and will justify this book's focus on a better understanding of 'how' these influences occur, instead of 'whether or not' these influences occur. Second, existing studies support the importance of understanding NGOs' work in the context of networks. Third, the resource-dependent nature of NGOs is important in understanding the way NGOs may operate within a network, as well as how they seek to ensure legitimacy in their activities. Fourth, the literature arguing the influence of regimes and collaboration suggests the importance of understanding the interactions between rules, norms, and the actors who interpret and apply them in various contexts. Finally, the review did not identify a sufficient body of literature which provides a comprehensive understanding of the influence of rules and norms on NGOs at different scales, including international, national, local, and within–NGO networks. Understanding these multiple influences of rules and norms can bring new aspects into the existing body of literature. The literature review has identified existing gaps and justifies the importance and unique contribution of this book, which examines the influence of rules and norms on the advocacy strategies of NGO coalitions.

Understanding the influence of rules and norms on actors' behaviours

This section identifies existing approaches used to analyse how rules and norms may influence NGOs' advocacy strategies.

New institutionalism

As discussed above, several scholars have adopted new institutionalism as a way of analysing how rules and norms may influence NGOs (Lawrence *et al.* 2002; Ohanyan 2012; Doh and Guay 2006; Ramanath 2009). This section analyses the theoretical approaches of new institutionalism and explores the possibilities of applying them to this book.

New institutionalism in political science emerged in the 1960s as an alternative explanation to behaviourism. Behaviourism attempted to explain political behaviour from an objective and quantitative perspective and through observable behaviour (Immergut 1998: 6; Hall and Taylor 1996: 936; John 1998: 57). The traditional (often referred to as old) institutional approach in political science, which dominated the discipline from the mid-nineteenth to the early twentieth century, focused primarily on the study of legal frameworks and administrative arrangements of government (Scott 2008: 6). Scholars of old institutionalism define an institution as an 'arena within which policy-making takes place' (John 1998: 38), which includes political organizations, administrative arrangements, and the laws and rules that play central roles in the policy process (John 1998: 38; Scott 2008: 6).

The proponents of new institutionalism rejected observable behaviour as a basis for understanding political behaviour, and proposed the importance of institutions in order to understand the behaviour of individuals (Immergut 1998: 6; Hall and Taylor 1996: 936). As compared with 'old' institutionalism, which focused on formal mechanisms such as law and governmental organizations, new institutionalism broadens its use of the term institution to also understand how institutions are shaped and how actors' behaviours are shaped by institutions (Roskin *et al.* 2008; Hall and Taylor 1996: 936).

New institutionalism gradually emerged from different disciplines, including sociology, economics, and political science, and expanded the definition of institution to include both formal and informal rules, norms, and even culture

(Hall and Taylor 1996). Political scientists, in their understanding of power and the behaviour of political actors, differentiated between different aspects of new institutionalism, including sociological institutionalism, institutional rational choice, and historical institutionalism (Hall and Taylor 1996). Accordingly, the definition of what is meant by 'institution' also varies. The primary difference lies between scholars who define an institution as 'organizational entities' and those who define an institution as 'rules, norms and strategies adopted by individuals operating within or across organizations' (Ostrom 1999: 37). Broadly speaking, all schools of new institutionalism take the latter approach to institution (Immergut 1998: 15).

Sociological institutionalism emerged out of a form of organization theory studied by sociologists (Hall and Taylor 1996: 946). As compared with the traditional theory of organizations, which assumed organizations are structured to maximize efficiency, sociological institutionalism takes the approach that organizations are structured not only to achieve efficiency, but primarily as a result of culturally specific practices (Hall and Taylor 1996: 14). Scholars of sociological institutionalism tend to include in their definition of institutions not only formal rules, procedures, or norms, but also 'symbol systems, cognitive scripts, and moral templates that provide the "frames of meaning" guiding human actions' (Hall and Taylor 1996: 947), breaking the conceptual divide between 'institution' and 'culture'. They interpret the relationships between institutional and individual behaviour based on their cultural and world views (cultural approach; Hall and Taylor 1996: 15). Two of the main proponents of sociological institutionalism are March and Olsen, who define political institutions as 'collections of interrelated rules and routines that define appropriate actions in terms of relations between roles and institutions', and claim that individuals act according to the rules and a logic of appropriateness of the institutions they enter into (March and Olsen 1989: 160). While this approach has a primary focus on understanding organizational behaviours, the logic of individual behaviour based on 'appropriateness' can be applied to NGO networks.

Another variation of new institutionalism is rational choice institutionalism, which draws analytical tools from new institutional economics (NIE) (Hall and Taylor 1996: 11). Proponents of NIE consider that 'institutions are humanly devised constraints that structure political, economic, and social interaction' (North 1991: 97). In addition, according to NIE, institutions are created to reduce transaction costs in society (Hall and Taylor 1996: 11). Rational choice institutionalism assumes the basic principle of actors' behaviour being rational; however, institutions also regulate the rational decisions of actors. Therefore actors do not act completely rationally (Kato 1996: 554).

The third type of new institutionalism is historical institutionalism, sometimes referred to as political institutionalism, which suggests that society is 'path-dependent', and attempts to understand how institutions produce society's 'path' (Hall and Taylor 1996: 9). Historical institutionalism understands institutions as rules, procedures, norms, and legacies that are a result of historical development (Immergut 1998: 15, 18). Historical institutionalism also seeks to understand how institutions shape ideas and influence policy, and how institutions influence the power relationships of groups (John 1998: 57). In understanding the relationship between institutional and individual behaviour, historical institutionalism takes both a rational and a cultural approach (Hall and Taylor 1996: 7–8).

The three forms of new institutionalism discussed in this section have different focuses and assumptions in understanding the relationships between institutions and individual behaviour. Two approaches to a fundamental understanding of human behaviour are the rational and the cultural approach (Hall and Taylor 1996). Rational choice institutionalists take a rational approach, whereas sociological institutionalists take a cultural approach, and historical institutionalists take both approaches. Some scholars criticize the approach taken by historical institutionalists, indicating that it does not provide a sophisticated understanding of how institutions affect individual behaviour (Hall and Taylor 1996: 17; Hay and Wincott 1998). However, by its nature, new institutionalism attempts to understand how institutions are adopted and accepted within a given society. New institutionalism attempts to integrate both rational and cultural approaches to human behaviour; as rational choice institutionalism in effect integrates a cultural approach as part of 'institutions', attempting to understand the rational behaviour of actors, taking into consideration institutional constraints. In essence, the divide between these two approaches is reduced.

The analytical lens provided by new institutionalism, which aims to understand how institutions influence actors' behaviours, is suited to the analytical focus of this book, which aims to understand how rules and norms influence actors. Understanding this influence is essentially a question of understanding how rules and norms interact with actors' behaviour. The following sections examine different approaches to understanding this interaction.

Typologies of relationships between formal and informal institutions

As discussed in the previous section, new institutionalists define institutions as the rules, norms, strategies, and culture adopted by a particular society (Immergut 1998: 15; Ostrom 1999: 37). Institutions may be formal or informal. Formal institutions are referred to as 'rules and procedures that are created, communicated, and enforced through channels widely accepted as official' (Helmke and Levitsky 2004). Informal institutions are referred to as institutions embedded in society, including customs, traditions, norms, and religions (Williamson 2000: 597).

Referring to the relationship between formal and informal institutions, Helmke and Levitsky argue that

> In comparative politics, the issue of how informal institutions sustain or reinforce – as opposed to undermine or distort – formal ones has not been

well researched. When institutions function effectively, we often assume that the formal rules are driving actors' behaviour. Yet in some cases, underlying informal norms do much of the enabling and constraining that we attribute to the formal rules.

(Helmke and Levitsky 2004: 734)

Based on this recognition, they suggest four typologies of relationships between formal and informal institutions, providing a useful analytical lens for understanding the institutional interactions (Helmke and Levitsky 2004: 723). As illustrated in Table 2.3, in a context where formal institutions are effective, informal institutions either complement or accommodate formal institutions. A complementary relationship occurs when informal institutions work to fill the gap not addressed by formal institutions (Helmke and Levitsky 2004: 728). Accommodating informal institutions are often created by actors who do not support the outcome generated by the application of formal rules, and attempt to reconcile their interests without violating formal institutions (Helmke and Levitsky 2004: 729).

In the context where formal institutions are ineffective, relationships between formal and informal institutions are either substitutive or competing. When informal institutions are applied to complement formal institutions by actors who seek outcomes aimed at by formal institutions, informal institutions play roles substituting for formal institutions. When formal institutions are not systematically enforced, informal institutions can be used by some actors to overpower the intended outcome of formal institutions, resulting in competing relationships between formal and informal institutions. Patrimonialism or corruption are typical examples of outcomes of this type of relationship (Helmke and Levitsky 2004: 729).

These typologies of relationships provide a useful means of understanding how rules and norms affect NGO actors. First, the typologies can provide an analytical lens to understand how both rules and norms affect certain actions taken by NGO coalitions. Second, as discussed by Helmke and Levitsky, different relationships between formal and informal institutions are created by the ways in which actors adopt them (Helmke and Levitsky 2004). In other words, understanding the interaction between formal institutions, informal institutions, and actors is necessary in order to understand the influence of institutions.

Table 2.3 A typology of informal institutions

Outcomes	Effective formal institutions	Ineffective formal institutions
Convergent	Complementary	Substitutive
Divergent	Accommodating	Competing

Source: Helmke and Levitsky (2004: 728).

Structure and agency

One of the important recent areas of debate in the social sciences, which has relevance for our understanding of the relationship between institutions and the behaviour of actors, is the relationship between structure and agency (Barnes 2003). One of the most widely known theories reflecting this relationship is the theory of structuration proposed by Anthony Giddens (Barnes 2003: 345). In this theory, Giddens claims that structure and agency are mutually dependent on each other, creating 'duality' (Giddens 1979: 69). This duality is further elaborated by Sewell's assertion that 'structures shape people's practices but it is also people's practices that constitute (and reproduce) structures' (Sewell 1992: 4). Giddens refers to structures as 'rules and resources that are organised as properties of social systems' (Giddens 1979: 66). Giddens also refers to structure as a social object that is external to human action, and 'a source of constraint on the free initiative of the independently constituted subject' (Giddens 1984: 16). Since Giddens' use of structure includes 'rules', structuration theory similarly supports the analytical focus of this book which considers that rules and norms shape actors' behaviours and *vice versa*.

Comparing structuration theory and theories of institution, Barley and Tolbert argue that both theories are complementary, sharing the premise that actions are shaped by institutions (Barley and Tolbert 1997: 112). Compared with theories of institutions, structuration theory focuses explicitly on the dynamics of how institutions are reproduced and created (Barley and Tolbert 1997: 112). Referring to the theory of structuration and the concept of duality, Barley and Tolbert provide a model that reflects how institutions influence action, and how actions reflect back to institutions (Barley and Tolbert 1997: 101).

Compared with the theory of institutions, which places more emphasis on how institutions shape actions, the theory of structuration focuses on the interaction between action and institutions. As discussed by Ohanyan in her analysis of network institutionalism, this aspect is important in understanding how NGOs' behaviours are shaped by the networks within which they operate, and *vice versa* (Ohanyan 2012).

The research undertaken in this book acknowledges the importance of taking the structure–agency relationship into consideration, and of understanding this duality, as agency could also influence structure, which later could influence agency. Structure–agency theory includes this iterative process of institutional evolution, which often occurs historically, and which can provide an important perspective for analysis of the influence of rules and norms.

In the context of understanding the influence of rules and norms over the strategies of NGO coalitions, structuration theory provides insights into understanding these relationships. While rules and norms affect NGOs, NGOs can also affect rules and norms. This relationship can occur at different levels: as an example, NGO coalitions may lobby to influence the development of national legislation associated with NGO operations in a particular country, which in return could affect the operation of NGOs. Another example, at

NGO coalitions' operational level, is that NGO coalitions may establish their own internal working rules which affect the way members operate within coalitions. While this book's main focus is on understanding how rules and norms affect NGO coalitions, it is important to keep in mind how NGO coalitions may also influence rules and norms, integrating the duality perspective of the structure–agency theory.

Institutional analysis and development framework

The institutional analysis and development (IAD) framework is one of the models developed by rational choice institutionalism scholars, which integrates the process of actors' collective choice shaped through both institutions and rationality (Ostrom 1999). It is a model which integrates various aspects of new institutionalism, and has been applied in many different contexts (Ostrom *et al.* 1994: 26). It is a framework for analysing 'how institutions affect the incentives confronting individuals and their resultant behaviour' (Ostrom 2005: 9). The framework was developed for analysing a wide variety of questions (Ostrom 2010b: 807). It was developed in order to integrate the work of various social science scholars including political scientists, economists, anthropologists, lawyers, sociologists, psychologists, and other scholars analysing institutions (Ostrom 2005: 8). Thus the IAD framework is compatible with multiple theories including economic theory, game theory, social choice theory, and theories of public goods and common-pool resources (Ostrom 2011: 8). The framework has been used to study various issues including privatization, macro-political systems, and public goods. However, as the framework was developed by scholars studying the use of natural resources, particularly of common pool resources, a considerable amount of academic work has used the IAD framework to study the problems of common pool resources and collective action among actors (Ostrom *et al.* 1994: 26; Buck 1999).

The IAD framework is illustrated in Figure 2.2, and includes three exogenous variables that affect actors' behaviours: 1) biophysical and material conditions; 2) attributes of community; and 3) rules-in-use. The framework contains general sets of variables suitable for examining human interactions in a diversity of institutional settings (Ostrom 2010a: 646). The key concept of the IAD framework is to illustrate how exogenous variables influence action situations (Ostrom 2010a: 646), where two or more actors are 'faced with a set of potential actions that jointly produce outcomes' (Ostrom 2005: 32).

Biophysical and material conditions are one of the exogenous variables that refer to the nature of resources affecting the action situations (Ostrom 2005: 22; McGinnis 2011: 174). Depending on whether resources have subtractability (actor A's consumption of the resource would affect the use by actor B) or levels of difficulty in excluding potential beneficiaries, resources are categorized into four types, as illustrated in Table 2.4 (McGinnis 2011: 174; Ostrom 2010a: 645). They are: 1) common-pool resources which have difficulty in excluding potential beneficiaries and also have high subtractability; 2) public goods where

Exogenous variables

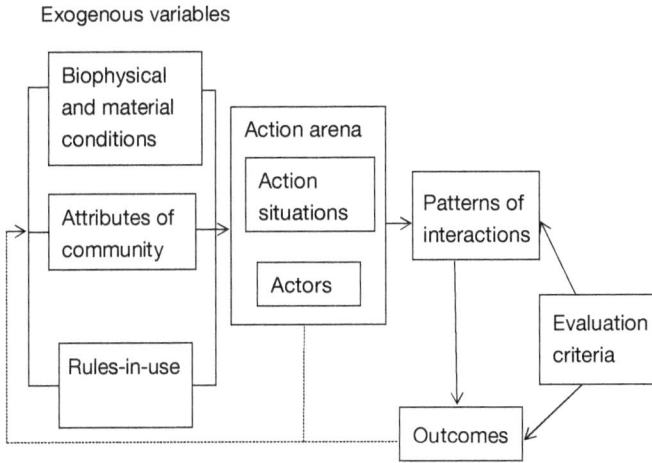

Figure 2.2 A framework for institutional analysis
Source: Ostrom (1999: 42; 2010a).

difficulty of excluding potential beneficiaries is high but subtractability is low; 3) private goods with low difficulty of excluding potential beneficiaries; and 4) toll goods with low difficulty in excluding potential beneficiaries and low subtractability (Ostrom 2010a: 645).

Table 2.4 Four types of goods

		Subtractability of use	
		High	*Low*
Difficulty of excluding potential beneficiaries	High	Common-pool resources (example: groundwater basins, lakes)	Public goods (example: peace and security of community, national defence)
	Low	Private goods (example: food, clothing, automobiles)	Toll goods (example: theatres, daycare centres)

Source: Adapted from Ostrom (2010a: 645, 2005: 24).

The second element of the exogenous variable is the attributes of the community. This variable includes 'norms of behaviour generally accepted in the community' or 'the levels of common understanding that potential participants share about the structure of a particular type of action arena'; the term 'culture' is often used to describe this variable (Ostrom 1999: 57). This may include the history of interactions, the extent of homogeneity/ heterogeneity of the community, and the size and composition of the community (Ostrom 2010a: 646; 2005: 27).

Rules-in-use refer to the working rules used by actors (Ostrom 2011: 19). Ostrom defines working rules as 'the set of rules to which participants would make reference if asked to explain and justify their actions to fellow participants' (Ostrom 1999: 51). Rules-in-use include a common understanding of who should and should not take certain actions (Ostrom 2010a: 647). The rules-in-use include both formal and informal rules and norms. These informal working rules may define how members of the community may operate according to 'logic of appropriateness' (March and Olsen 1989), or in some cases, activities to be 'embedded' (Granovetter 1985) within the socio-cultural context of a particular society.

The action arena is the social space where individuals interact and activities take place (Ostrom 1999: 42). It is a dependent variable in the context of the IAD framework (Ostrom 2005: 16). The action arena consists of two components: actors and action situations (Ostrom 2005: 13). The action situation is referred to as a 'situation when two or more individuals are faced with a set of potential actions that jointly produce outcomes' (Ostrom 2005: 32). Patterns of interactions are created as a result of an action situation, which result in an outcome. Some of the activities feed back into exogenous variables.

Existing studies that applied the IAD framework found it to be general enough to allow its application across a diverse range of issues and levels (Smajgl *et al.* 2009: 150). The framework was also found to be useful for analysing rules and how they affect behaviours of individuals and organizations (Imperial 1999). Another study found that the IAD framework provides more levels of detail compared with other existing institutional frameworks (Ghorbani *et al.* 2012: 96). Some scholars have also attempted to use the IAD framework to explain the role of NGOs in conflict resolution (Nesbit 2003), or civil society actors in their struggle against hydropower dams (Myint 2005). Carlsson describes policy networks as collective action, and suggests the use of the IAD framework as a way to analyse policy networks (Carlsson 2000). He values the IAD framework for the analysis of policy networks as it provides a framework for context sensitive analysis (Carlsson 2000: 513). While the IAD framework is certainly a useful framework for understanding the institutional factors affecting the advocacy strategies of NGO coalitions, examples of use of the framework for this purpose are still limited to date, constituting a gap in its application.

The IAD framework provides a useful basis for the analysis in this book, for several reasons. First, the IAD framework aims to explain how institutions affect the behaviours of actors, particularly with its focus on how rules affect behaviours (Ostrom 2005: 9). Second, the framework provides an explanation of different elements affecting a community, and an analysis of their interactions in adopting these elements. This approach supports the analysis of how interactions occur among rules, norms, and actors. Finally, the framework can be applied by different scholars using a variety of theories (Ostrom 1999: 40). While the IAD framework bases its fundamental principle on rational choice institutionalism, it also accommodates aspects of other varieties of new institutionalism, including historical aspects and cultural appropriateness. This flexibility in the framework allows researchers to accommodate and adjust the inclusion of factors most suitable for the specific research objective.

Conclusion

This chapter reviews and analyses the literature discussing advocacy strategies, factors influencing NGOs, and rules and norms, and their influence over actors' behaviours. The analysis clarifies four key findings which support the development of the analytical approach developed for the case studies in this book.

The first finding is that the existing literature has already discussed that rules and norms affect actors' behaviour, including NGOs' behaviour (Lawrence *et al.* 2002; Ho and Edmonds 2008; Brinkerhoff 1999; Bryant 2001). However, an understanding of how this influence occurs is less developed. This finding justifies the starting point of this book, which assumes that rules and norms influence actors' behaviour. Therefore this book focuses on examining *how* rules and norms influence the advocacy strategies of NGO coalitions, without first asking whether or not they *do* influence them.

The second finding is that an understanding of the influence of rules and norms requires an understanding of how actors interact and adopt rules and norms in their actions (Bryant 2001; Helmke and Levitsky 2004; Ostrom 2005). This approach is prevalent in the IAD framework, and the study of relationships between formal and informal institutions by Helmke and Levitsky (2004). The third finding is that in understanding networks, it is important to identify relationships and resource dependency among actors (Keck and Sikkink 1999; Jordan and van Tuijl 2000; Rhodes 1997). This relationship is also important in understanding how rules and norms are adopted by network members. The second and third findings emphasize the importance of understanding interactions, a key concept in this book's analysis. Finally, new institutionalism provides a most relevant focus for better understanding of the influence of rules and norms on actors' behaviours.

Based on this analysis of the existing frameworks and theories used to study the influence of rules and norms on behaviour, the next chapter develops the methodology used in the analysis of the cases discussed in this book.

References

Abercrombie, Nicholas, Stephen Hill, and Bryan S. Turner. 2000. *The Penguin Dictionary of Sociology*, 4th edn. London: Penguin Books.
Allen, Robert, ed. 2000. *The Penguin Dictionary*. London: Penguin Books.
Almond, Gabriel A. Jr, G. Bingham Powell, Kaare Strøm, and Russel J. Dalton. 2004. *Comparative Politics Today: A World View*, 8th edn. London: Pearson Longman.
Anderson, Alison. 1997. *Media, Culture and Environment*. London: UCL Press.
Barley, Stephen R., and Pamela S. Tolbert. 1997. Institutionalization and structuration: studying the links between action and institution. *Organization Studies* 18 (1): 93.
Barnes, Barry. 2003. The macro/micro problem and the problem of structure and agency. In *Handbook of Social Theory*, edited by G. Ritzer and B. Smart. London: Sage.
Black, Max. 1962. *Models and Metaphors: Studies in Language and Philosophy*. Ithaca: Cornell University Press.

Bodansky, Daniel. 1999. The legitimacy of international governance: a coming challenge for international environmental law? *American Journal of International Law* 93 (3): 596–624.

Brinkerhoff, Derick W. 1999. State–civil society networks for policy implementation in developing countries. *Review of Policy Research* 16 (1): 123–147.

Brown, Lesley, ed. 1993. *The New Shorter Oxford English Dictionary on Historical Principles*. Oxford: Clarendon Press.

Bryant, Raymond L. 2001. Explaining state–environmental NGO relations in the Philippines and Indonesia. *Singapore Journal of Tropical Geography* 22 (1): 15–37.

Buck, Susan J. 1999. Multiple-use commons, collective action, and platforms for resource use negotiation. *Agriculture and Human Values* 16 (3): 237–239.

Bunn, Colin, and Victoria Ayer. 2004. *Working with the Media*. Advocacy Expert Series Module 3. Phnom Penh: PACT Cambodia. www.k4health.org/sites/default/files/advocacy_series_module3.pdf. Accessed 7 June 2015.

Büsgen, M. 2006. *NGOs and the Search for Chinese Civil Society Environmental Non-Governmental Organisations in the Nujiang Campaign*. Working Paper Series No. 422. The Hague: Institute of Social Studies. www.eu-china.net/upload/pdf/materialien/michael%20buesgen-diplom_08-08-19.pdf. Accessed 7 June 2015.

Callon, M. 2001. Actor network theory. In *International Encyclopedia of the Social & Behavioral Sciences*, edited by N. J. Smelser and P. B. Baltes. Oxford: Elsevier Science.

Carlsson, Lars. 2000. Policy networks as collective action. *Policy Studies Journal* 28 (3): 502–520.

Charnovitz, Steve. 1997. Two centuries of participation: NGOs and international governance. *Michigan Journal of International Law* 18: 183–286.

Cohen, Jean L., and Andrew Arato. 1994. *Civil Society and Political Theory*. Cambridge: MIT Press.

Covey, Jane G. 1995. Accountability and effectiveness of NGO policy alliances. *IDR Reports* 11 (8). www.hapinternational.org/pool/files/covey-effectiveness-and-accountability.pdf. Accessed 19 February 2011.

Croteau, David, and William Hoynes. 2003. *Media Society: Industries, Images, and Audiences*. 3rd edn. London: Pine Forge Press.

Curzon, L. B. 1998. *Dictionary of Law*, 5th edn. London: Financial Times Professional.

De Santo, Elizabeth M. 2010. 'Whose science?' Precaution and power-play in European marine environmental decision-making. *Marine Policy* 34 (3): 414–420.

DiMaggio, Paul J., and Walter W. Powell. 1983. The iron cage revisited: institutional isomorphism and collective rationality in organizational fields. *American Sociological Review* 48 (2): 147–160.

Doh, Jonathan P., and Terrence R. Guay. 2006. Corporate social responsibility, public policy, and NGO activism in Europe and the United States: an institutional-stakeholder perspective. *Journal of Management Studies* 43 (1): 47–73.

Dolwick, Jim. 2009. 'The social' and beyond: introducing actor-network theory. *Journal of Maritime Archaeology* 4 (1): 21–49.

Edwards, Michael. 2004. *Civil Society*. Cambridge: Polity Press.

Finnemore, Martha, and Kathryn Sikkink. 1998. International norm dynamics and political change. *International Organization* 52: 887–917.

Fisher, Julie. 1998. *Nongovernments: NGOs and Political Development of the Third World*. West Hartford, CT: Kumarian Press.

Fox, Jonathan. 2010. Coalitions and networks. In *International Encyclopedia of Civil Society. Volume 1*, edited by K. H. Anheier, S. Toepler, and R. A. List. New York: Springer.

Fox, Leslie M., and Priya Helweg. 1997. *Advocacy Strategies for Civil Society: A Conceptual Framework and Practitioner's Guide*. A report prepared for USAID. www.innonet.org/resources/files/Advocacy_Strategies_for_Civil_Society.pdf. Accessed 18 April 2013.

Gamson, William A. 2004. Bystanders, public opinion, and the media. In *The Blackwell Companion to Social Movements*, edited by D. A. Snow, S. A. Soule, and H. Kriesi. Oxford: Blackwell Publishing.

Garner, Bryan A., ed. 2004. *Black's Law Dictionary*, 8th edn. Eagan, MN: Thomson West.

Gemmill, Barbara, and Abimbola Bamidei-Izu. 2002. The role of NGOs and civil society in global environmental governance. In *Global Environmental Governance*, edited by D. C. Etsy and M. H. Ivanova. New Haven, CT: Yale School of Forestry & Environmental Studies.

Ghorbani, Amineh, Virginia Dignum, and Gerard Dijkema. 2012. An analysis and design framework for agent-based social simulation. Paper presented at AAMAS (Autonomous Agents and Multi-Agent Systems) 2011 Workshops, Taipei, Taiwan, May 2011.

Giddens, Anthony. 1979. *Central Problems in Social Theory: Action, Structure and Contradiction in Social Analysis*. London: Macmillan.

—— 1984. *The Constitution of Society: Outline of the Theory of Structuration*. Cambridge: Polity Press.

Gieryn, Thomas F. 1983. Boundary-work and the demarcation of science from non-science: strains and interests in professional ideologies of scientists. *American Sociological Review* 48 (6): 781–795.

Gordon, Graham. 2002. *Advocacy Toolkit: Practical Action in Advocacy*. Roots resources 2. Teddington, UK: Tearfund. http://tilz.tearfund.org/Publications/ROOTS/Advocacy+toolkit.htm. Accessed 23 April 2013. Original version no longer available online; updated version at http://tilz.tearfund.org/~/media/files/tilz/publications/roots/english/advocacy%20toolkit/second%20edition/tearfundadvocacytoolkit.pdf.

Graber, Doris A. 2002. *Mass Media & American Politics*, 6th edn. Washington DC: CQ Press.

Granovetter, Mark. 1985. Economic action and social structure: the problem of embeddedness. *American Journal of Sociology* 91 (3): 481–510.

Gu, H. 2008. Participatory citizenship and sustainable development: redefining 'public' in contemporary China and Japan. Paper presented at 17th Biennial Conference of the Asian Studies Association of Australia, Melbourne, 1–3 July 2008.

Guston, David H. 2001. Boundary organizations in environmental policy and science: an introduction. *Science, Technology, & Human Values* 26 (4): 399–408.

Haas, Peter M. 1992. Introduction: epistemic communities and international policy coordination. *International Organization* 46 (1): 1–35.

—— 2007. Epistemic communities. In *The Oxford Handbook of International Environmental Law*, edited by D. Bodansky, J. Brunnee, and E. Hey. Oxford: Oxford University Press.

Haggard, Stephan, and Beth A. Simmons. 1987. Theories of international regimes. *International Organization* 41 (3): 491–517.

Hall, Peter, and Rosemary C. R. Taylor. 1996. Political science and the three new institutionalisms. *Political Studies* XLIV: 936–957.

Hay, Colin, and Daniel Wincott. 1998. Structure, agency and historical institutionalism. *Political Studies* 46 (5): 951–957.

Helmke, G., and S. Levitsky. 2004. Informal institutions and comparative politics: a research agenda. *Perspectives on Politics* 2 (4): 725–740.

Hill, Michael. 2009. *The Public Policy Process*, 5th edn. London: Pearson Education.

Ho, P., and R. L. Edmonds, eds. 2008. *China's Embedded Activism: Opportunities and Constraints of a Social Movement*. Routledge Studies on China in Transition. London: Routledge.

Hogwood, Brian. 1995. The 'growth' of quangos: evidence and explanations. *Parliamentary Affairs* 48 (2): 207–225.

Hsu, Carolyn. 2010. Beyond civil society: an organizational perspective on state–NGO relations in the People's Republic of China. *Journal of Civil Society* 6 (3): 259–277.

Hudson, Alan. 2001. NGOs' transnational advocacy networks: from 'legitimacy' to 'political responsibility'? *Global Networks* 1 (4): 331–352.

Huitema, Dave, and Esther Turnhout. 2009. Working at the science–policy interface: a discursive analysis of boundary work at the Netherlands Environmental Assessment Agency. *Environmental Politics* 18 (4): 576–594.

Immergut, Ellen M. 1998. The theoretical core of the new institutionalism. *Politics & Society* 26 (1): 5–34.

Imperial, M. T. 1999. Institutional analysis and ecosystem-based management: the institutional analysis and development framework. *Environmental Management* 24 (4): 449–465.

John, Peter. 1998. *Analysing Public Policy*. London: Continuum.

Jordan, Lisa, and Peter van Tuijl. 2000. Political responsibility in transnational NGO advocacy. *World Development* 28 (12): 2051–2065.

Kato, Junko. 1996. Institutions and rationality in politics – three varieties of neo-institutionalists. *British Journal of Political Science* 26 (4): 553–582.

Keck, Margaret E., and Kathryn Sikkink. 1998. *Activists Beyond Borders: Advocacy Networks in International Politics*. Ithaca/London: Cornell University Press.

—— 1999. Transnational advocacy networks in international and regional politics. *International Social Science Journal* 51 (159): 89–101.

Koh, Harold Hongju. 1998. 1998 Frankel Lecture: Bringing international law home. *Houston Law Review* 35.

Lawrence, Thomas B., Cynthia Hardy, and Nelson Phillips. 2002. Institutional effects of interorganizational collaboration: the emergence of proto-institutions. *Academy of Management Journal* 45 (1): 281–290.

Lenschow, Andrea. 1997. Variation in EC environmental policy integration: agency push within complex institutional structures. *Journal of European Public Policy* 4 (1): 109–127.

Lister, Sarah. 2003. NGO Legitimacy: Technical Issue or Social Construct? *Critique of Anthropology* 23 (2):175–192.

Lovan, W. Robert, Michael Murray and Ron Shaffer. 2004. Participatory governance in a changing world. In *Participatory Governance*, edited by R. Lovan, W. M. Michael, and S. Ron. Farnham: Ashgate.

Lowndes, Vivien. 1996. Varieties of new institutionalism: a critical appraisal. *Public Administration* 74 (2): 181–197.

March, James G., and Johan P. Olsen. 1989. *Rediscovering Institutions: The organizational Basis of Politics*. New York: Free Press.

Martin, Elizabeth, and Jonathan Law, eds. 2006. *Oxford Dictionary of Law*, 6th edn. Oxford: Oxford University Press.

McCombs, Maxwell E., and Donald L. Shaw. 1972. The agenda-setting function of mass media. *Public Opinion Quarterly* 36 (2): 176–187.

McGinnis, Michael D. 2011. An introduction to IAD and the language of the Ostrom workshop: a simple guide to a complex framework. *Policy Studies Journal* 39 (1): 169–183.

McQuail, Denis. 1987. *Mass Communication Theory: An Introduction*, 2nd edn. London: Sage.

Meierotto, Lisa. 2009. The uneven geographies of transnational advocacy: the case of the Talo Dam. *Journal of Environmental Management* 90 (S3): S279–S285.

Meyer, Thomas. 2002. *Media Democracy: How the Media Colonize Politics*. Cambridge: Polity Press.

Myint, Tun. 2005. Strength of 'weak' forces in multilayer environmental governance: cases from the Mekong and Rhine River Basins. PhD thesis, School of Public and Environmental Affairs and School of Law, Indiana University, Bloomington, IN.

Naim, Moises. 2007. What is a GONGO?: how government-sponsored groups masquerade as civil society. *FP*, www.foreignpolicy.com/articles/2007/04/18/what_is_a_gongo.

Negrine, Ralph. 1994. *Politics and the Mass Media in Britain*, 2nd edn. London: Routledge.

Nesbit, B. 2003. The Role of NGOs in conflict resolution in Africa: an institutional analysis. Paper presented at Institutional Analysis and Development Mini-Conference, Indiana University, Bloomington, IN, USA, 3–5 May.

Newman, M. E. J. 2010. *Networks: An Introduction*. Oxford: Oxford University Press.

North, Douglass C. 1991. Institutions. *Journal of Economic Perspectives* 5 (1): 97–112.

Ohanyan, Anna. 2012. Network institutionalism and NGO studies. *International Studies Perspectives* 13 (4): 366–389.

Ostrom, Elinor. 1999. Institutional rational choice: an assessment of the institutional analysis and development framework. In *Theories of the Policy Process*, edited by P. A. Sabatier. Oxford: Westview Press.

—— 2005. *Understanding Institutional Diversity*. Princeton: Princeton University Press.

—— 2010a. Beyond markets and states: polycentric governance of complex economic systems. *American Economic Review* 100 (3): 641–672.

—— 2010b. Response: the institutional analysis and development framework and the commons. *Cornell Law Review* 95: 807.

—— 2011. Background on the institutional analysis and development framework. *Policy Studies Journal* 39 (1): 7–27.

Ostrom, Elinor, Roy Gardner, and James Walker. 1994. *Rules, Games, & Common-Pool Resources*. Ann Arbor, MI: University of Michigan Press.

Parsons, Wayne. 1995. *Public Policy: An Introduction to the Theory and Practice of Policy Analysis*. Aldershot, UK: Edward Elgar.

Pfeffer, Jeffrey, and Gerald R. Salancik. 2003. *External Control of Organizations: A Resource Dependence Perspective*. Palo Alto, CA: Stanford University Press.

Pielke, Roger Jr, and Roberta A. Klein, eds. 2010. *Presidential Science Advisors: Perspectives and Reflections on Science, Policy and Politics*. London: Springer.

Prell, Christina. 2012. *Social Network Analysis: History, Theory & Methodology*. London: Sage.

Ramanath, Ramya. 2009. Limits to institutional isomorphism: examining internal processes in NGO–government interactions. *Nonprofit and Voluntary Sector Quarterly* 38 (1): 51–76.

Randall, Vicky. 1993. The media and democratisation in the Third World. *Third World Quarterly* 14 (3): 625–646.

Rhodes, R. A. W. 1996. The new governance: governing without government. *Political Studies* XLIV: 652–667.

—— 1997. *Understanding Governance*. Buckingham, UK: Open University Press.

—— 2007. Understanding governance: ten years on. *Organization Studies* 28 (08): 1243–1264.

Roskin, Michael G., Robert L. Cord, James A. Medeiros, and Walter S. Jones. 2008. *Political Science: An Introduction*. Upper Saddle River, NJ: Pearson Prentice Hall.

Sabatier, Paul A., and Hank C. Jenkins-Smith. 1999. The advocacy coalition framework: an assessment. In *Theories of Policy Process*, edited by P. A. Sabatier. Oxford: Westview Press.

Saxon-Harrold, Susan K. 1990. Competition, resources, and strategy in the British nonprofit sector. In *The Third Sector: Comparative Studies of Nonprofit Organizations*, edited by H. Anheier and W. Seibel. Berlin: Walter de Gruyter & Co.

Scheufele, D. A. 1999. Framing as a theory of media effects. *Journal of Communication* 49 (1): 103–122.

Scott, John, and Gordon Marshall, eds. 2005. *Oxford Dictionary of Sociology*. Oxford: Oxford University Press.

Scott, W. Richard. 2008. *Institutions and Organizations: Ideas and Interests*, 3rd edn. London: Sage.

Sewell, William H. Jr. 1992. A theory of structure: duality, agency, and transformation. *American Journal of Sociology* 98 (1): 1–29.

Sikkink, Kathryn. 2005. Patterns of dynamic multilevel governance and the insider–outsider coalition. In *Transnational Protest & Global Activism*, edited by D. Della Porta and S. Tarrow. Oxford: Rowman & Littlefield.

Simon, Adam F., and Jennifer Jerit. 2007. Toward a theory relating political discourse, media, and public opinion. *Journal of Communication* 57 (2): 254–271.

Smajgl, Alex, Anne Leitch, and Tim Lynam. 2009. *Outback Institutions: An Application of the Institutional Analysis and Development (IAD) Framework to Four Case Studies in Australia's Outback*. DKCRC Report 31. www.nintione.com.au/resource/DKCRC-Report-31-Outback-Institutions_Application-of-the-IAD-framework.pdf. Accessed 13 December 2013.

Spitzer, Robert J. 1993. Defining the media–policy link. In *Media and Public Policy*, edited by R. J. Spitzer. London: Praeger.

Sprechmann, Sofia, and Emily Pelton. 2001. *Advocacy Tools and Guidelines: Promoting Policy Change*. Atlanta, GA: CARE. www.care.org/getinvolved/advocacy/tools.asp. Accessed 18 April 2013.

Steinberg, Richard, and Walter W. Powell. 2006. Introduction. In *The Nonprofit Sector: A Research Handbook*, 2nd edn, edited by W. W. Powell and R. Steinberg. New Haven/London: Yale University Press.

Suchman, Mark C. 1995. Managing legitimacy: strategic and institutional approaches. *Academy of Management Review* 20 (3): 571–610.

Tarrow, Sidney. 1998. *Power in Movement: Social Movement and Contentious Politics*, 2nd edn. Cambridge: Cambridge University Press.

Turnhout, Esther, Matthijs Hisschemöller, and Herman Eijsackers. 2007. Ecological indicators: between the two fires of science and policy. *Ecological Indicators* 7: 215–228.

UNICEF. 2010. Advocacy toolkit: a guide to influencing decisions that improve children's lives. New York: UNICEF. www.unicef.org/evaluation/files/Advocacy_Toolkit.pdf. Accessed 18 April 2013.

Van der Heijden, Hein-Anton. 1997. Political opportunity structure and the institutionalisation of the environmental movement. *Environmental Politics* 6 (4): 25–50.

Van der Sluijs, Jeroen. 2005. Uncertainty as a monster in the science–policy interface: four coping strategies. *Water Science & Technology* 52 (6): 87–92.

Wasserman, Stanley, and Katherine Faust. 1994. *Social Network Analysis: Methods and Applications*. New York: Cambridge University Press.

Weible, Christopher M., Paul A. Sabatier, and Kelly McQueen. 2009. Themes and variations: taking stock of the advocacy coalition framework. *Policy Studies Journal* 37 (1): 121–140.

Wells-Dang, Andrew. 2011. Informal pathbreakers: civil society networks in China and Vietnam. PhD thesis, Department of Politics & International Studies, School of Government and Society, University of Birmingham.

Williamson, Oliver E. 2000. The new institutional economics: taking stock, looking ahead. *Journal of Economic Literature* 38 (3): 595–613.

World Association of Non-Governmental Organizations. 2010. *Overview of NGOs*. www.ngohandbook.org/index.php?title=Overview_of_NGOs. Accessed 4 May 2014.

World Bank. 1998. *Operational Manual. GP 14.70 – Involving Nongovernmental Organizations in Bank-Supported Activities*. http://web.worldbank.org/WBSITE/EXTERNAL/PROJECTS/EXTPOLICIES/EXTOPMANUAL/0,,contentMDK:20064711~menuP K:4564189~pagePK:64709096~piPK:64709108~theSitePK:502184,00.html. Accessed 4 May 2014.

Wu, Fengshi. 2003. Environmental GONGO Autonomy: unintended consequences of state strategies in China. *Good Society* 12 (1): 35–45.

Yanacopulos, Helen. 2005. The strategies that bind: NGO coalitions and their influence. *Global Networks* 5 (1): 93–110.

Young, Doug. 2012. *The Party Line: How the Media Dictates Public Opinion in Modern China*. Somerset, NJ: Wiley.

3 Framework for analysis

NGO coalitions along the Mekong River

Introduction

Based on the overview in Chapter 2, this chapter identifies the research methodology used for this book. The chapter first discusses a framework for analysis, which will be applied to analyse case studies. NGO coalitions' strategies associated with the Xayaburi hydropower dam on the Mekong River are discussed as comparative cases. The chapter also presents data collection and analysis strategies, and concludes with an illustration of how the analytical framework is used to structure this book.

Analytical framework

Chapter 2 suggested that new institutionalism is a theoretical approach that is well suited for the analysis used in this book. The analytical framework developed here combines various approaches and concepts identified in that review. The institutional analysis and development (IAD) framework was identified in Chapter 2 as a useful basis for developing this book's analytical framework, due to its adaptability to different purposes, the inclusion of key components necessary for the analysis, and its focus on understanding the interaction among rules, norms, actors, and other factors. The IAD framework is therefore used as a point of departure and modified to meet the needs of this book. The revised and developed analytical framework is presented in Figure 3.1.

One modification is based on the findings from the literature that identifies the importance of understanding the interactions among rules, norms, and actors. This aspect was identified through the literature discussing relationships between formal and informal institutions (Helmke and Levitsky 2004), structuration theory (Giddens 1984), and the literature on NGOs discussing the influence of institutions (Lawrence *et al.* 2002). The original IAD framework focuses only on one-way relationships (albeit in a circular loop) as institutions are considered to be exogenous variables explaining action situations. The modification of the framework was made so that there is now a two-way relationship where rules and norms affect actors, but actors also affect rules and norms.

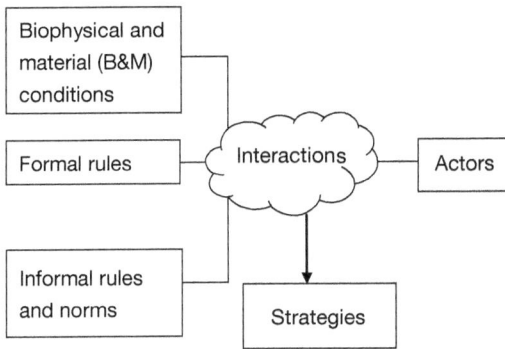

Figure 3.1 Analytical framework of this book

Second, the relationships and interactions among actors were also identified in the literature as an important factor determining how rules and norms affect actors (Helmke and Levitsky 2004). The original IAD framework integrates this point through its emphasis on action situations, where two or more sets of actors interact within an action arena. Ostrom admits that it has created confusion to include all 'action arenas', 'action situations', and 'actors' in the framework, as actors are already an integrated part of action situations that take place within an action arena (Ostrom 2011). As 'arena' is not specifically identified in the literature as an important factor, but rather the 'interaction of actors', this book's framework focuses on actors and their interactions.

Exogenous variables have been modified to suit the needs of this book, which is to understand how rules and norms affect strategies. The book divides these concepts into 'formal rules' and 'informal rules and norms'. The original terms 'attributes of communities' and 'rules in use' are replaced by these two key concepts. 'Biophysical and material conditions' (B&M conditions) remain in the framework, as this factor is an important aspect in the case study analysis of the use of water resources in the Mekong River. Here a resource perspective becomes an important factor that affects actors' behaviours.

The component 'outcome' from the original IAD framework is replaced by 'strategies' in order to provide specific meanings to the outcomes applicable in this analysis. The feedback arrow from outcome to exogenous variables, as well as 'evaluation criteria', are removed from the framework as they are beyond the scope of this book.

Case studies: NGO coalitions along the Mekong River

The analytical framework developed in the previous section was used to analyse how rules and norms influence the advocacy strategies of NGO coalitions. A comparative case-study method is used as this research strategy is considered suitable to answer 'why' and 'how' questions. The method also enables the researcher to deal with linkages that need to be explained or traced over time, and is suitable for conducting in–depth analysis (Yin 2003; Sarantakos 2005: 212). In order to examine how different rules and norms influence the advocacy strategies of NGO coalitions, a comparative case analysis was conducted of NGO coalitions that are active in the same issue, while operating under different rules and norms. By comparing cases, the research aims to analyse and develop explanations – which also constitutes the methodological core in comparative politics (Almond *et al.* 2004: 31).

As a research context, the advocacy work of NGO coalitions associated with the hydropower development plan within the Mekong River was chosen. More specifically, the book studies the advocacy strategies of NGO coalitions attempting to influence the involved states' decision-making during the Procedures for Notification, Prior Consultation and Agreement (PNPCA) process for the Xayaburi hydropower dam, planned at the Laos' geographical location along the Mekong River. This research context was selected for several reasons. First, as the Mekong River is a transboundary river, the impact from the Xayaburi hydropower dam in Laos is expected to occur within other downstream countries, notably Thailand, Cambodia, and Vietnam. The transboundary nature of the project stimulated the activities of NGO coalitions in all the affected riparian countries, providing an opportunity to compare the actions of different NGO coalitions operating under different rules and norms, yet working on the same issue. Second, the Xayaburi hydropower dam is the first project on the Mekong River that utilized the PNPCA process, one of the procedural rules under the Agreement on the Cooperation for the Sustainable Development of the Mekong River Basin (1995 Mekong Agreement). The 1995 Mekong Agreement has been the legal regime governing the Mekong River for the past 20 years, and the findings from this book have the potential to provide input into the future governance of the Mekong River under the 1995 Mekong Agreement regime. Considering a large number of similar development projects proposed within the Mekong River Basin, the findings are expected to have practical application beyond the case of the Xayaburi dam. Finally, the study on the PNPCA process with reference to the Xayaburi hydropower dam is also important from an academic point of view, as it was the first case where prior consultation requirement of the PNPCA was applied.

As comparative cases, NGO coalitions operating in Cambodia and Vietnam were chosen for several reasons. The main rationale for choosing these two countries is the comparative approach known as 'method of difference', which compares two similar cases that differ with respect to the variables that are the subject of the study (Hopkin 2010: 291). In the context of the Xayaburi hydropower dam, both Cambodia and Vietnam are located downstream of the dam, and neither country would benefit from this hydropower development. The main beneficiaries are Thailand and Laos, as the Xayaburi dam is located in Laos and constructed with investment from Thailand, and most electricity generated by it would be used in Thailand (MRC Secretariat 2011: 10). Second, Cambodia and Vietnam represent two different types of political system which exist among the four riparian countries that are involved in the decision-making process for the Xayaburi dam: Cambodia and Thailand are democracies, whereas Laos and Vietnam are communist countries ruled by one political party. Comparing cases operating within different political systems highlights these differences, particularly in formal rules, which are important factors shaping political systems. This comparison also allows a certain level of generalization of similar networks within the Mekong region, as the two cases represent the two major political systems existing within the region (Sarantakos 2005: 98). Third, existing studies on NGO networks in Cambodia and Vietnam are limited, compared with those in Thailand, which has a longer history of NGO and civil society movements on dams. Laos was not selected as a case study as NGO activity, particularly on hydropower dams, was limited. However, the study aims to bring insights to the linkages between rules and norms and advocacy strategies of NGO coalitions from neighbouring countries, which could be of interest to the future development of NGOs in Laos.

During the PNPCA process, a coalition of NGOs and civil society actors was active in both Cambodia and Vietnam: in Cambodia, the coalition is known as the Rivers Coalition in Cambodia (RCC), and in Vietnam it is the Vietnam Rivers Network (VRN). At the time of the field research conducted for this book, no other major domestic NGO networks were active during the PNPCA process. Therefore these two networks were chosen as case studies for this book. The period of advocacy activities studied is from September 2010 until August 2012. This time frame was chosen for several reasons. September 2010 is the date when Laos informed the Mekong River Commission (MRC) of its intension to build the Xayaburi dam, instigating the PNPCA process. During the following two years, various activities were conducted both by state and non-state actors. This provided a sufficient amount of information to conduct case studies.

Collection and analysis of data

The data collection and analysis strategy involves three key methods: 1) desktop document analysis and literature review; 2) interviews and field observations; and 3) text analysis using a grounded theory approach. This three-pronged approach also allows triangulation in the analysis of the data. The overall approach is illustrated in Figure 3.2, and the analysis using these three methods aimed at identifying relationships between rules, norms, and the advocacy strategies of NGO coalitions. The following section discusses in detail how these three methods were applied in the research.

Desktop document search and literature review

A desktop document search and literature review were conducted throughout the research in order to identify variables related to each element of the analytical framework and to provide background information and understanding

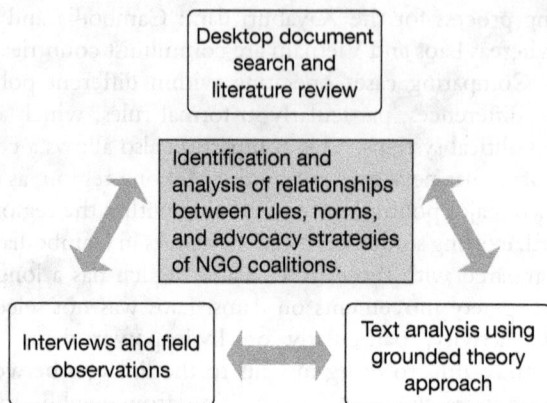

Figure 3.2 Approach to data analysis

of the rules, norms, and context of the case. A backwards-and-forwards process between literature and other forms of data is part of the grounded theory approach described in more detail below. A categorical analysis of some of these documents was used as one method to identify important variables. Categorical document analysis is a systematic analysis of documents based on emerging categories identified by the researcher (Sarantakos 2005: 294), and this method was used to identify potential rules and norms that might have the ability to influence NGO coalitions' behaviours. This process of ongoing document analysis was conducted throughout the research.

Interviews and field observations

Semi-structured interviews were conducted with key informants during two periods of fieldwork in the Mekong region. The analysis of these interviews further identified potential relationships between rules, norms, and NGO coalitions' strategies. The interviews were conducted with the aim of obtaining an understanding of: 1) strategies adopted by the case study coalitions during the Xayaburi hydropower dam PNPCA process; and 2) key informants' perceptions of how rules and norms affected advocacy strategies of NGO coalitions over the Xayaburi hydropower dam during the PNPCA process. Furthermore, they were used to: 3) obtain supplementary information about rules, norms, actors, and advocacy strategies of NGO coalitions relevant to the case studies. Prior to the interviews, indicative questions were developed to help guide the interview process.

The interviews were conducted with key informants who had an understanding of the social context and with NGOs in the case study areas. In order to gain a balanced view of the research subject, interviews were carried out with informants who were both members and non-members of the NGO coalitions. Interviewees were selected initially through identification of the key individuals engaged in the NGO coalitions. The author's existing contacts with NGO actors in the region facilitated access to these actors. Further interviewees were identified using the snowball sampling method, where the interviewer asks interviewees to suggest other people to be interviewed on the subject matter (Sarantakos 2005: 165). Most interviews were conducted during the author's field visits in the Mekong region in November 2011, and in June–August 2012. A total of 72 informants were interviewed, most once, and some twice.

Three categories of interviewees were identified: those familiar primarily with case study issues in Cambodia (35 interviewees, three interviewed twice); those familiar primarily with case study issues in Vietnam (26 interviewees, three interviewed twice); and those familiar with case studies in the Mekong regional context (11 interviewees, two interviewed twice). Interviews are cited anonymously in this book; Cambodian interviews are indicated with the letter C followed by an interviewee number, Vietnamese interviewees are cited with the letter V followed by an interviewee number, and regional interviewees are cited with the letter R followed by an interviewee number.

In addition to the interviews, the research benefited from observations made by the author. There were two types of observation. First, those that took place during the fieldwork, where the author had the opportunity to attend a total of seven workshops and meetings of NGO and civil society actors on associated topics. In addition, a field visit to the Xayaburi dam site was conducted in November 2011 in order to gain first-hand understanding of the research context. The second type of observation was based on the author's experience of living and working with NGOs and international organizations in Southeast Asia, particularly in the Mekong region, for seven years from 2002–09. This initial period of observation equipped the author with a basic understanding about the societies, and the rules and norms existing in the case study region. Information obtained through these observations enhanced the author's understanding and analysis. A list of interviews and observed meetings is included in the Annex.

Text analysis using grounded theory approach

As the main focus of this book is in understanding the relationship between rules, norms, and advocacy strategies of NGO coalitions, its scientific ontology is based on constructionism, which considers reality as being constructed; and its epistemology on interpretivism, which takes the approach that meaning is constructed through interpretation of individuals (Sarantakos 2005: 37–40). This ontology and epistemology provides a basis for qualitative analysis, and the rationale for the use of interviews and the researcher's interpretation as a way of understanding the relationships between rules, norms, and advocacy strategies of NGO coalitions.

Field data, including transcriptions and interview notes, were analysed through careful readings and identification of the recurring themes and relationships among rules, norms, and advocacy strategies of the case study coalitions. This method of text analysis builds on grounded theory, which allows researchers to identify recurring themes through analysis of empirical data (Tischer *et al.* 2000: 76–77; Glaser and Holton 2004). Grounded theory is a useful approach for the purpose of this book, as it facilitates analysis and identification of the relationships between rules, norms, and advocacy strategies of NGO coalitions. However, the approach adopted for this book does not follow grounded theory in its strict sense, as grounded theory requires researchers to analyse the data without any preconceptions (Grix 2004: 111). Since the research focused on identifying relationships between rules, norms, and advocacy strategies by establishing key criteria and a study objective in advance, it did not strictly follow the coding procedure of grounded theory (Corbin and Strauss 1990; Holton 2007: 266). However, coding was used as a tool to assist in identifying key concepts and themes linking rules, norms, and advocacy strategies. ATLAS Ti research software supporting text analysis was used to code interview data to identify recurring themes.

The main sources of data included documents and interviews which were used in an in-depth qualitative case study (Yin 2003: 86). These were collected through desktop reviews and fieldwork. As the PNPCA process had already taken place prior to the author's fieldwork, opportunities for direct observation of this process were limited. However, during the fieldwork, the research took advantage of any opportunities that arose for direct observation that could assist in a better understanding of the characteristics and operational context of the NGO networks. The key steps in data collection and analysis are summarized in Table 3.1.

Table 3.1 Steps of analysis

Step	Activity	Source of data	Method
1	Initial identification of formal rules, informal rules, and norms	Law, policy, and formal rules in Mekong region, Cambodia, Vietnam, and within NGO coalitions	Desktop search and literature review
		Literature on political culture	
2	Identification of key actors involved in the Xayaburi dam, their roles, responsibilities, and relationships	Organizational websites, literature on the Mekong River and the Xayaburi dam	Desktop search and literature review
3	Identification of biophysical and material conditions	Literature on the Mekong River and the Xayaburi dam	Desktop search and literature review
4	Identification of strategies and tactics adopted by the case study coalitions, and their background	Organizational websites, Interviewees	Desktop search Interviews Observation Literature review

Step	Activity	Source of data	Method
5	Analysis of interview data to identify recurring themes and key concepts linking rules, norms, and advocacy strategies	Interview records (notes and transcripts) and any additional data collected during the fieldwork	Text analysis using grounded theory methods
	Further identification of relevant formal and informal rules and norms		
6	Review of literature associated with key themes	Literature	Literature review
7	Analysis of interaction between rules, norms, and actors	All the data above	Final integrated analysis

Conclusion

This chapter presents the approach employed in this book. Based on the theoretical overview discussed in Chapter 2, new institutionalism was identified as the theoretical ground for the book. The IAD framework was adopted as a basis to develop the analytical framework, modified through integrating key insights from the literature review. The chapter also discusses the data collection and data analysis strategy. The use of comparative case studies was selected to highlight how different rules and norms affect advocacy strategies of NGO coalitions working over the same hydropower dam issues. The data analysis methodology was based on three types of analytical method: desktop document search and literature review, interviews and field observations, and text analysis based on grounded theory. This research methodology, particularly the analytical framework, shapes the structure of the book. Figure 3.3 illustrates how key components of the analytical framework are discussed in each chapter.

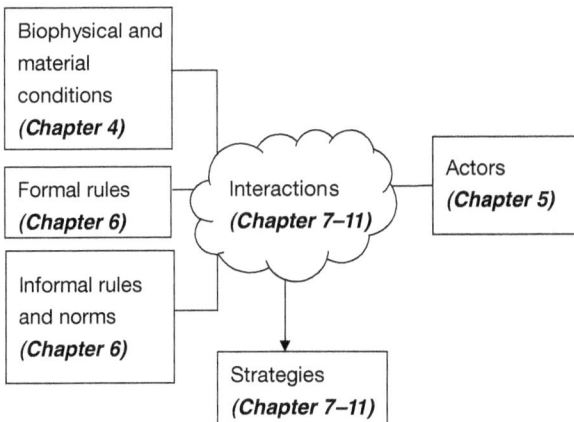

Figure 3.3 Structure of the book based on the analytical framework

In Part II of this book, three key components that are considered to affect advocacy strategies are discussed: biophysical and material conditions (Chapter 4), actors (Chapter 5), and rules and norms (Chapter 6). Each chapter discusses the components that are applicable for both Cambodian and Vietnamese case studies. Two remaining components (strategies and interactions) are discussed in Part III (chapters 7–11).

References

Almond, Gabriel A. Jr, G. Bingham Powell, Kaare Strøm, and Russel J. Dalton. 2004. *Comparative Politics Today: A World View*, 8th edn. London: Pearson Longman.

Corbin, Juliet, and Anselm Strauss. 1990. Grounded theory research: procedures, canons, and evaluative criteria. *Qualitative Sociology* 13 (1): 3.

Giddens, Anthony. 1984. *The Constitution of Society: Outline of the Theory of Structuration.* Cambridge: Polity Press.

Glaser, B. G., and J. Holton. 2004. Remodeling grounded theory. *Forum: Qualitative Social Research* 5 (2).

Grix, Jonathan. 2004. *The Foundations of Research*. New York: Palgrave.

Helmke, G., and S. Levitsky. 2004. Informal institutions and comparative politics: a research agenda. *Perspectives on Politics* 2 (4): 725–740.

Holton, Judith A. 2007. The coding process and its challenges. In *The Sage Handbook of Grounded Theory*, edited by A. Bryant and K. Charmaz. London: Sage.

Hopkin, Jonathan. 2010. The comparative method. In *Theory and Methods in Political Science*, 3rd edn, edited by D. Marsh and G. Stoker. New York: Palgrave Macmillan.

Lawrence, Thomas B., Cynthia Hardy, and Nelson Phillips. 2002. Institutional effects of interorganizational collaboration: the emergence of proto-institutions. *Academy of Management Journal* 45 (1): 281–290.

MRC Secretariat. 2011. *Proposed Xayaburi Dam Project – Mekong River: Prior Consultation Project Review Report*. www.mrcmekong.org/assets/Publications/Reports/PC-Proj-Review-Report-Xaiyaburi-24-3-11.pdf. Accessed 25 October 2011.

Ostrom, Elinor. 2011. Background on the institutional analysis and development framework. *Policy Studies Journal* 39 (1): 7–27.

Sarantakos, Sotirios. 2005. *Social Research*, 3rd edn. New York: Palgrave Macmillan.

Tischer, Stefan, Michael Meyer, Ruth Wodak, and Eva Vetter. 2000. *Methods of Text and Discourse Analysis*. London: Sage.

Yin, Robert K. 2003. *Case Study Research: Design and Methods*, 3rd edn. London: Sage.

Part II

Biophysical and material conditions, actors, rules, and norms

4 Biophysical and material conditions

Introduction

The biophysical and material conditions associated with the case study is a key component of the analytical framework introduced in Chapter 3. This book examines the advocacy strategies of the Rivers Coalition in Cambodia and the Vietnam Rivers Network over the Xayaburi hydropower dam on the Mekong River. Therefore this chapter first introduces a general resource perspective on the Mekong River, followed by a description of hydropower dam development and future hydropower potential on the Mekong River. The chapter also provides an introduction to the Xayaburi hydropower dam, and its potential impact on Cambodia and Vietnam. The chapter concludes with a discussion of key findings.

Resource perspective of the Mekong River

The Mekong River is the longest river in Southeast Asia, and the tenth largest river in the world, with a total drainage area of 795,000 km² and an approximate length of 4,900 km (MRC undated). The river originates in the Tibetan Plateau of China, flowing through Myanmar (formerly called Burma until 1989), Laos, Thailand, and Cambodia, until finally entering the sea in the Mekong Delta in Vietnam (MRC undated). The river provides important sources of livelihood for the riparian populations; there are 60 million people living in the Lower Mekong Basin alone (MRC 2010b: 3). The Mekong's fishery is one of the largest freshwater fisheries in the world, with an estimated value of US$3.9–7.0 billion annually (MRC 2010b: 13). In Cambodia, where one of the case studies of this book is located, the fisheries sector provided employment to approximately six million people in 2007, including people engaged in fisheries on a part-time basis (Joffre et al. 2012: 1). Eighty per cent of animal protein intake by Cambodians comes from fish (Joffre et al. 2012: 1; Hortle 2007: xi; C7 2012). Agriculture is one of the most important economic activities in the Lower Mekong Basin, totalling ten million hectares for rice production (MRC 2010b: 8).

Agriculture, particularly in the lower part of the river, depends on fertile soils enriched with sediment from the Mekong River (MRC 2010b: 8). The Mekong Delta is considered the 'rice bowl' of Vietnam, and is one of the most important and fertile regions of the country. Irrigation is one of the largest water uses in the Lower Mekong Basin (LMB), which includes Cambodia, Thailand, Laos and Vietnam (MRC 2010b: 9). The Mekong River is also home to a wide diversity of species, and to some globally threatened species such as the Irrawaddy dolphin, Siamese crocodiles and giant catfish (MWBP 2005). The estimated number of fish species living in the Mekong River varies between 758 and 1500 depending on the report (MRC 2003: 5; Baran *et al.* 2007: 12; Orr *et al.* 2012: 926; WWF 2013). The large number of species indicates the importance of the Mekong from the perspective of freshwater biodiversity.

Hydropower development on the Mekong River: historical perspectives

The large volume of water associated with the river has attracted hydropower dam engineers for many years. The estimated potential hydropower capacity of the whole basin is in the order of 50,000–64,750 megawatts (MW), of which 30,000 MW is within Cambodia, Thailand, Laos and Vietnam, comprising the LMB (MRC 2010b: 20; ICEM 2010b: 11). Seventy per cent of the hydropower potential within the LMB is located in Laos (MRC 2010b: 20). The Mekong River and surrounding natural resources have also attracted the interest of non-riparian nations, particularly from the West. France, for example, was one of the first nations which sent their explorers to this river through the work of the Mekong Exploration Commission, as early as 1866 (Keay 2005: 289). The potential for tapping into this resource and political power struggles among nation states are closely linked to the governance of the river and to plans for developing hydropower dams. This historical perspective is an important factor that has shaped current resource use and the interests of different actors, affecting the resource perspective of the case study.

The modern attempt to manage the use of the Mekong River commenced in the 1950s, with the support of the United Nations and the 'Western bloc' states during the Cold War, particularly the United States (Bakker 1999: 213). The Mekong Committee was established in 1957, consisting of the four lower Mekong states: Cambodia, Thailand, Laos, and Vietnam (Jacobs 1995). The Committee was established through facilitation and support by the United Nations Economic Commission for Asia and the Far East (ECAFE; later renamed the United Nations Economic and Social Commission for Asia and the Pacific) (Jacobs 1995: 138). Two riparian states did not take part in the early Mekong Committee: China was not a member of the ECAFE at the time, which precluded its participation, and Burma was not interested in participating (Jacobs 1995: 139).

The focus of the Committee was to plan for the socio-economic development of the Lower Mekong countries through large-scale infrastructure projects such as hydropower, irrigation, and flood control projects (Jacobs 1995: 139). The cascade of dams on the mainstream was planned during this time (Hirsch 2010: 313). The United States was the largest non-riparian donor to the Committee during this period, which coincided with the early stages of the Cold War. Providing a blueprint for lifting the Lower Mekong countries out of poverty through hydropower development was one of the political strategies that the United States adopted in order to attract the Mekong countries into the Western bloc (Jacobs 1995: 139; Hirsch 2010: 313; Middleton *et al.* 2009: 26).

The work of the Committee and the development of hydropower dams faced difficulty during the period of political instability in the region in the 1970s (Jacobs 1995). Laos experienced a civil war from 1953–75 which reflected a fight between the Vietnamese-backed communist group Pathet Lao and the United States-backed monarchy (Stuart-Fox 1997). Cambodia also faced internal conflict during the 1970s and 1980s. The brutality of the Cambodian conflict reached its peak when the Khmer Rouge took over power in 1975, forcing the majority of Cambodian citizens into collective agricultural labour in severe conditions, and killing thousands of their own people (Vickery 1984). During the same period, Vietnam was a major battlefield of the Cold War, with fighting continuing for 20 years until the mid-1970s (Allen 2008). These political instabilities hindered the participation of riparian countries in the Committee, and neither Cambodia, Vietnam, nor Laos appointed members to the Committee between 1976 and 1977 (Jacobs 1995: 143). Though Laos and Vietnam resumed their participation in the Committee in 1978, the Khmer Rouge-controlled Cambodian government did not join the Committee. This lack of full membership had a significant impact on development projects related to the Mekong River. The Joint Declaration of Principles for Utilization of the waters of the Lower Mekong Basin, adopted by the member states of the Committee, did not allow unilateral appropriation of mainstream water without prior approval by the other basin states (Article X, *Mekong Joint Declaration* 1975; Jacobs 1995: 143). In addition, the deteriorating political situation in the region contributed to the decline of funding for the Committee (Jacobs 1995; Middleton *et al.* 2009: 8). As an attempt to resume the work of the Committee, the Interim Mekong Committee was established in 1978 with Thailand, Vietnam, and Laos as members. However, the operation of the Interim Committee struggled due to even lower levels of funding (Jacobs 1995: 143).

While hydropower projects on the mainstream were therefore halted due to the political instability in the region, projects on the tributaries progressed in the 1980s and early 1990s. In Thailand, the Pak Mun dam on the Mun River, one of the tributaries of the Mekong, was completed in 1994 with partial funding from the World Bank (Nippanon *et al.* 2000). The construction of the Yali falls dam in Vietnam started in 1993 on the Se San River, another tributary of the Mekong; the dam is located approximately 70–80 km upstream from the Cambodian border (Hirsch 2010; International Rivers 2002). In Laos, several tributary dams were also built in the 1990s with partial or full funding from the Asian Development Bank (ADB), such as the Xeset dam, the Nam Leuk dam, and the Theun Hinboun dam (Hirsch 2010). In addition, the Nam Theun 2 dam in Laos, which is the largest of all the tributary dams with a capacity of 1107 MW, financed by the World Bank, created a decade of controversy among proponents and opponents of such developments (Hirsch 2010: 316). Despite such controversy the dam was finally built and commenced operation in 2010 (Nam Theun 2 Power Company undated).

Political stability in the Mekong region was regained with the signing of the Paris Peace Agreement in 1991, which ended decades of internal conflict in Cambodia. The Cold War also ended in the early 1990s with the collapse of the Soviet Union (Middleton *et al.* 2009: 27; Bakker 1999: 214). As political stability returned to the region, the international community revived its efforts to bring the basin countries together for the management of the Mekong River. With the facilitation of the United Nations Development Programme (UNDP), the Mekong Working Group was established in 1993 among Cambodia, Laos, Thailand, and Vietnam in order to conduct preparatory work on the formulation of a new framework for cooperation within the Mekong riparian nations (Mekong Secretariat 1994: 6). As a result of this work, the four LMB riparian countries signed the Agreement on the Cooperation for the Sustainable Development of the Mekong River Basin in 1995 (1995 Mekong Agreement). The Agreement established the Mekong River Commission (MRC), which is an institutional framework for the member states to cooperate (Article 11 *Mekong Agreement* 1995). The institutional framework under the 1995 Mekong Agreement is discussed further in Chapter 5.

As peace and stability were regained in the region and a new governance regime for the Mekong River was established, the western bilateral aid agencies, the World Bank, and the ADB reactivated their interest in investment and aid opportunities (Middleton *et al.* 2009: 28). The development of hydropower dams, particularly on the Mekong's mainstream, was a priority of these international donors (Middleton *et al.* 2009: 28). With financial support from the French government and the UNDP, the Mekong Secretariat conducted a study investigating the possibilities for run-of-the-river hydropower projects on the mainstream of the Mekong River in 1994 (Mekong Secretariat 1994: 31). The Xayaburi hydropower dam, which is the main context of this book's case study, was identified as a potential project in the Mekong Secretariat's study (TEAM Consulting Engineering and Management Co. Ltd. 2008: Section 1-1).

The end of political conflict in the region also brought opportunities for economic development through regional integration. As a major framework for regional economic development, the Greater Mekong Sub-region (GMS) programme was launched by the ADB and was endorsed by the region's governments in 1992. The GMS programme emphasized the physical interconnectivity of the region, proposing major infrastructure projects including hydropower dams and regional transmission lines (Middleton *et al.* 2009: 28; ADB 2012).

By the time the MRC was established, increased criticism of the impacts of a large-scale hydropower dam was highlighted globally, creating additional hurdles for dam developers in the implementation of their projects. In 1997, the World Bank and the International Union for Conservation of Nature (IUCN) initiated a global stakeholder dialogue on large dams, resulting in the establishment of the World Commission on Dams (WCD) in 1998 (Moore *et al.* 2010). In 2000, the WCD conducted a review of the impacts from large dams globally, and produced a set of recommendations to minimize negative impacts (WCD 2000). The Pak Mun dam, built on the tributary of the Mekong River in Thailand, was used as a case study by the WCD. The study pointed out the economic inefficiency of the dam and the negative impacts on livelihoods, indicating that the dam should not have proceeded in its current setting if a proper assessment had been conducted (Nippanon *et al.* 2000: iv–xiv).

Civil society movements against large hydropower dams also became strong in Thailand in the 1980s (Jacobs 2002). Five major dam projects were subject to public protest, and only one of them was built, the Pak Mun dam (Bakker 1999: 216) – and the Pak Mun dam faced strong public protest. This resulted in the altering of its operation and the opening of the dam gate for four months a year to allow fisheries to continue, which had a negative impact on its output capacity (Foran 2006; Mekong Watch 2002).

By the late 1990s, there was a sentiment of wariness towards large-scale dam development among donors to the MRC, and their support to the Commission's work diverted to environment and fisheries (Hirsch 2010: 317). In addition, the Asian financial crisis in 1997 affected the economy of the region, and hit Thailand in particular. This in turn reduced the need for new sources of electricity (Middleton *et al.* 2009: 32). Therefore the development of hydropower dams on the Mekong appeared to no longer be a priority. However, as the region's economy recovered in the late 2000s, the plans for development of mainstream hydropower dams reappeared on the agenda (Hirsch 2010: 317). This time the plan was picked up by relatively 'new' players in the region: private-sector hydropower developers primarily from Thailand, Vietnam, China, and Malaysia (Middleton *et al.* 2009: 23). As of 2010, there were 12 hydropower project plans within the mainstream of the LMB. Ten of them were located within the territory of, or on the border of, Laos, and the rest of the planned construction was located in Cambodia (ICEM 2010b: 30–31).

Responding to growing concerns over the social and environmental impacts of hydropower dams, together with increased interest in the development of hydropower dams on the Mekong, the MRC Council approved a change to the existing MRC hydropower programme in 2005, and the Initiative on Sustainable Hydropower (ISH) commenced operation in 2008 (MRC 2006a: 30; 2008: 25). At the same time, the MRC started to collaborate with the ADB and the World Wide Fund for Nature (WWF), an international environmental NGO, in formulating environmental criteria for hydropower development (MRC 2006b: 42). Headed by a new Chief Executive Officer whose previous experience included working at the secretariat of the WCD, the ISH conducted a series of multi-stakeholder consultation dialogues on hydropower dams in 2008 (MRC 2008). These dialogues aimed to engage various actors in the debate over hydropower dam development.

In 2008 the decision was made to conduct a strategic environmental assessment (SEA) of hydropower dam plans on the mainstream of the Mekong River (MRC 2008: 26). The SEA study was commissioned by the MRC, and conducted by the International Centre for Environmental Management (ICEM), an international consultancy firm based in Hanoi (ICEM 2010b). The SEA report, which was finalized in 2010, recommended a ten-year deferral of mainstream hydropower dams until further studies on their impact could be conducted (ICEM 2010b). The SEA report became the source of much debate on hydropower dam issues within the region, and its recommendations were used as a reference point by NGOs, as well as by international donors, journalists, and some MRC member countries, in their arguments related to the Xayaburi hydropower dam (Socialist Republic of Viet Nam 2011; Save the Mekong 2011; MRC 2011a). According to the SEA report, as of 2010, 12 hydropower dams were planned in the Lower Mekong Basin (ICEM 2010b: 27), and the Xayaburi dam was the one of them.

The Xayaburi hydropower dam

The Xayaburi hydropower dam site is located on the mainstream of the Mekong River within the territory of Laos. The site is located near Ban Pakneun and borders two provinces, Xayaburi province and Luang Prabang province (TEAM Consulting Engineering and Management Co. Ltd. 2008: 2-2). Located in the northern part of Laos, the town of Luang Prabang is an old capital city, and is designated as a UNESCO World Heritage Site (UNESCO 2013). The dam site is located approximately 100 km downstream of Luang Prabang town (Ministry of Energy and Mines 2012). Figure 4.1 illustrates the location of the dam. The Xayaburi dam is one of the dams listed as potential mainstream hydropower dams in the Mekong's mainstream run-of-river hydropower study, conducted by the Mekong Secretariat in 1994 (ICEM 2010b: 27; TEAM Consulting Engineering and Management Co. Ltd. 2008: 1-1).

As of 2010, 12 hydropower dams were planned on the mainstream of the LMB, and the Xayaburi dam was one out of ten located within Laos' territory and borders (ICEM 2010a: 6). For Laos, the least developed country within the Mekong region, development through hydropower is one of its major poverty reduction strategies, as it aims to become the 'battery' of Southeast Asia (Ministry of Planning and Investment 2011: 125). Laos' seventh five-year national socio-economic development plan (2011–15) therefore indicates development of large-scale hydropower stations along the Mekong River and its tributaries as one of the key strategies to promote industrialization and modernization (Ministry of Planning and Investment 2011: 188).

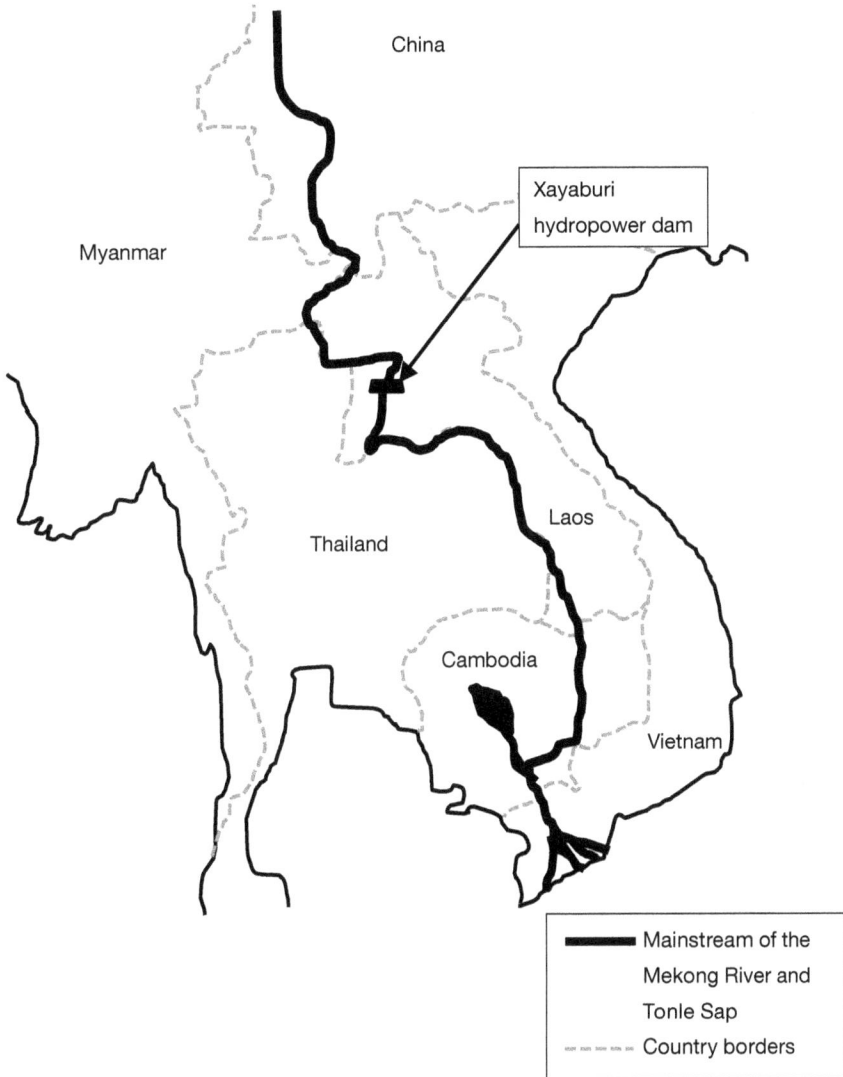

Figure 4.1 Location of the Xayaburi hydropower dam

Map created by the author from TEAM Consulting Engineering and Management Co. Ltd. (2008); Geospatial Information Authority of Japan (undated); Joo (2006)

Placing a physical obstacle in a flowing river creates a number of serious impacts on a river's ecosystem, and the Xayaburi dam is no exception. The dam is expected to result in multiple impacts on downstream communities and countries in Laos, Thailand, Cambodia, and Vietnam. One of the expected key impacts of the dam is on fisheries. The Mekong River is home to an estimated 758–1500 fish species (WWF 2013; Baran *et al.* 2007: 12; Orr *et al.* 2012: 926;

MRC 2003: 5) and most species are highly migratory. According to the fisheries expert group commissioned by the MRC during a prior consultation project review for the Xayaburi dam, migration of 23 to 100 fish species is considered to be potentially affected (MRC Secretariat 2011: 92). This number includes five species listed on the IUCN Red List of Threatened Species (MRC Secretariat 2011: 92).

Another key impact from the dam is the sediment and associated geomorphological impact downstream (Graf 2006; MRC Secretariat 2011: Annex 3). The river water naturally carries certain levels of sediment. When a dam is placed on a river, creating a reservoir, the sediment accumulates at the bottom of the reservoir unless flushed out from the dam. The excessive trapping of sediment in the reservoir creates a 'hungry water' phenomenon, which is the disequilibrium between the water and sediment discharges. This imbalance in the sediment load downstream of the river can create geomorphological changes in the landscape. The water that flows out from a reservoir contains less sediment compared with that of the original river water. This 'sediment-hungry' water, which flows downstream in a river system, creates bank erosion due to a shortage of coarse materials in the water. The sediment also carries nutrients such as phosphorus and nitrogen. The creation of a reservoir within a river can cause trapping of the nutrients attached to sediments. Accumulation of nutrients and a reduction of turbidity due to the creation of a reservoir can also increase algal growth, which affects the water quality (MRC Secretariat 2011: Annex 3).

The Mekong River transports sediment and associated nutrients downstream through its regular flow and seasonal flood, and the agricultural practices within the Lower Mekong Basin rely on this fertile soil transported through its natural flow system. Sediment is particularly important for agriculture and the landscape in the Mekong Delta. Despite the importance of sediment, there is a lack of baseline data on sediment in the Mekong, which creates uncertainty for some of the estimates made in the technical review (MRC Secretariat 2011: 93).

The scope of the environmental impact assessment (EIA) for the Xayaburi hydropower dam was limited to the watershed area of the proposed site, ten kilometres downstream from the barrage site, and impoundment area (TEAM Consulting Engineering and Management Co. Ltd. 2010: 1–2). During the PNPCA of the Xayaburi dam, no transboundary impact study was conducted, which was the main concern of downstream riparian countries, particularly Cambodia and Vietnam (Kingdom of Cambodia 2011; Socialist Republic of Viet Nam 2011). The MRC Council in December 2011 agreed to conduct a comprehensive impact study for mainstream hydropower dams (MRC 2012). However, the process for proceeding with this study has been rather slow, creating frustration among some of the parties involved (MRC 2013a, 2013b, 2013c). In particular, Cambodia and Vietnam are the two riparian countries that do not receive direct benefits from the Xayaburi dam, yet will be impacted by the development.

Potential impact of the Xayaburi hydropower dam

Cambodia

With regard to the Xayaburi hydropower dam, Cambodia has two contradicting interests. On one hand, Cambodia is located downstream of the Xayaburi hydropower dam. This physical location creates the key concern influencing the Cambodian government's position, the potential impact on Cambodian fisheries, upon which the livelihoods of approximately six million Cambodians depend (C7 2012; Foster 2011; Joffre *et al.* 2012: 1). Impact is expected particularly in six provinces located along the Mekong River's mainstream (see Figure 4.2 for the locations) including Stung Treng province, Kratie province, Kampong Cham province, Prey Veng province, Kandal province, Takeo province, and the capital Phnom Penh (MRC 2011b: 10). In addition, the changes in flow may potentially affect provinces along the Tonle Sap Lake, where the Mekong River reverses flow during the wet season (MRC 2010a: 30; MRC Secretariat 2011: 45). Tonle Sap Lake is the largest freshwater lake in Cambodia, where over one million people depend on the natural resources of the lake for their livelihoods. The reverse flow from the Mekong River accounts for 52 per cent of the inflow to the lake (Kummu and Sarkkula 2008: 186).

On the other hand, Cambodia itself has two hydropower dam investments planned on the mainstream of the Mekong River: the Sambor dam and the Stung Treng dam. The Sambor dam, which is planned in Kratie province, will have a total capacity of 2600 MW, and the Stung Treng dam, planned in Stung Treng province, will have a total capacity of 980 MW (ICEM 2010b: 31). Both dams would lie in the north-eastern part of Cambodia within the stretch of the Mekong River between Khon Falls in Laos down to Kratie province in Cambodia. This area is an important destination for fish migration from the Tonle Sap Lake, and building the Sambor dam and the Stung Treng dam will disrupt this important fish migration route (ICEM 2010b: 95). According to the SEA for 12 hydropower dams planned on the Mekong's mainstream of the Lower Mekong Basin, these are two out of the three dams estimated to cause the largest impacts on fish production (ICEM 2010b: 103).

As a consequence, Cambodia has two contradictory interests with respect to the Xayaburi dam. On one hand, the dam may have a significant impact on Cambodia's fishery. On the other hand, if Cambodia opposes Laos' attempt to build the Xayaburi dam, there is a risk that Cambodia's hydropower dam projects on the Mekong River's mainstream will face similar objections from other riparian states. These physical and material conditions of the Xayaburi dam constitute the context in which governments and NGOs of the Lower Mekong approached the issue of the Xayaburi dam.

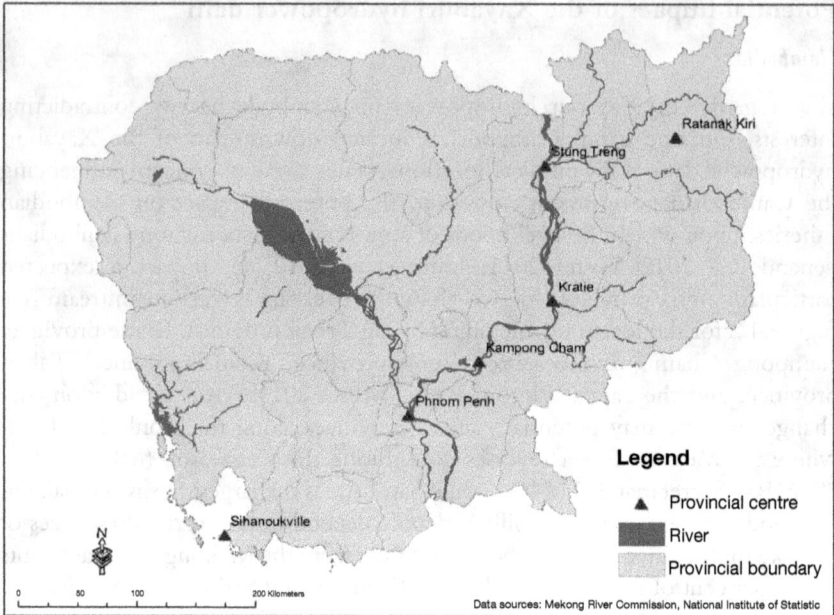

Figure 4.2 Map of Cambodian provinces and the Mekong River
Map created by Sopheak Chann

Vietnam

Vietnam is situated where the main impact from the dam is expected to occur, in the Mekong Delta (see Figure 4.3), and the Xayaburi dam is expected to affect the sediment flow of the river, changing the river's geomorphology (MRC Secretariat 2011: 59). The Delta, the rice bowl of Vietnam, plays a vital role in the Vietnamese economy (Quang 2002). According to a soil expert with knowledge of soil in the Mekong Delta, the Delta already faces geomorphological changes due to the construction of the Chinese dams (V18 2012). Changes upstream also affect nutrient transports in the river water which are particularly important for agricultural production in the Mekong Delta (MRC Secretariat 2011: 45). The main expected impacts of the Xayaburi dam on the Delta are through the loss of sediment through increased erosion (MRC Secretariat 2011; MRC 2010a: 35). These physical and material conditions played an important role in determining the Vietnamese government's position on the Xayaburi dam, which is discussed in the following sections.

While Vietnam is the country furthest downstream in the context of the Xayaburi hydropower dam, it is also an upstream country for Cambodia, with some of the tributaries of the Mekong River such as the Se San and the Sre Pok rivers originating in Vietnam. The Vietnamese government still invests in hydropower development upstream in these tributaries (Grimsditch 2012), and these hydropower investments impact the downstream communities in

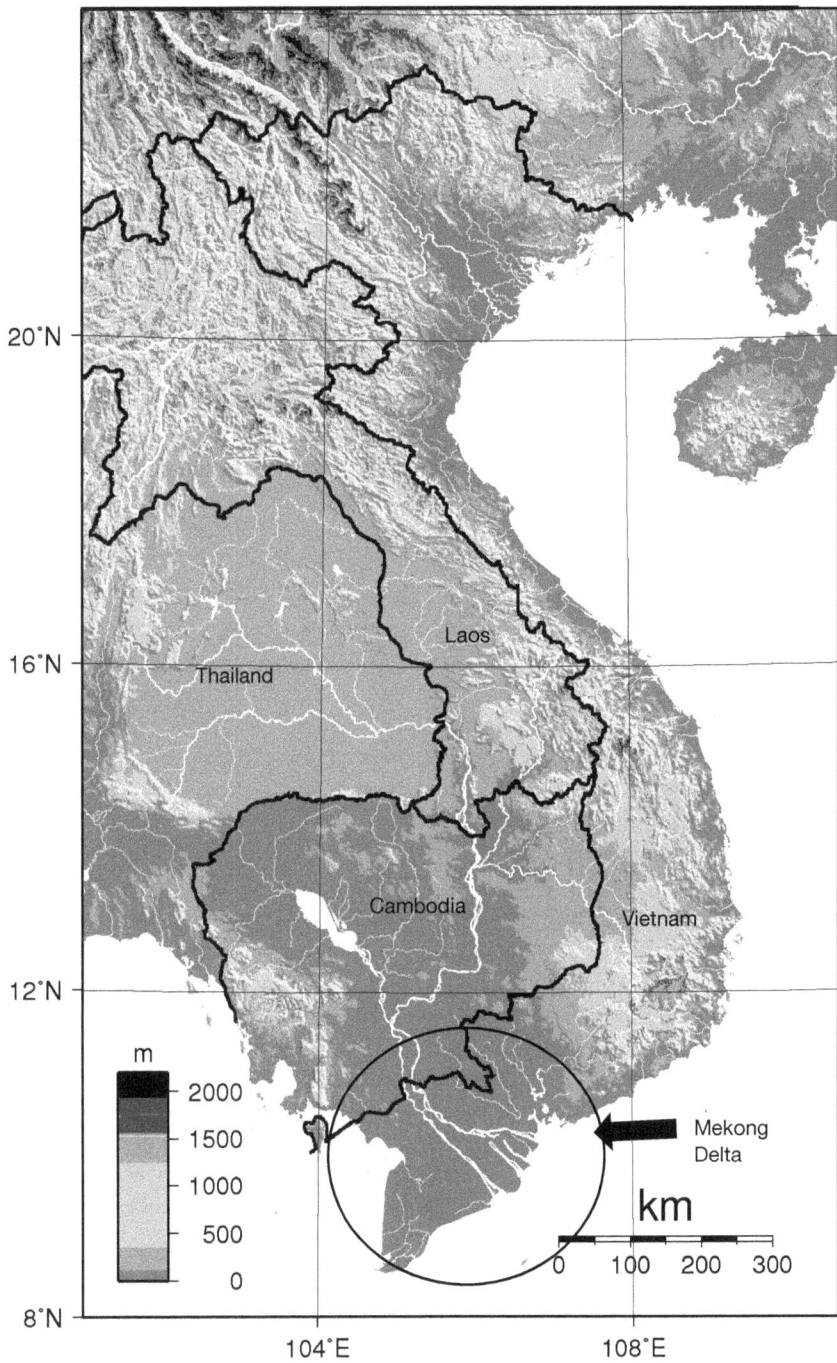

Figure 4.3 Location of the Mekong Delta

Source: Sadalmelik (2007)

Cambodia through increased flooding incidences, fluctuation in river water, changes in water quality, and reduction in fishery catches (Wyatt and Baird 2007). Some interviewees claimed that the potential impacts from these tributary dams can be as significant as those of the mainstream dams (C27 2012). This can potentially have an impact on the Mekong Delta, which is located further downstream.

NGOs and Cambodian communities affected by the existing dams on these tributaries have voiced their concerns over the impact of dams in the past. However, prior consultation was not conducted for these dams as presumably it was not considered necessary under the terms of the 1995 Mekong Agreement for tributary dams (Article 5 *Mekong Agreement* 1995). The 1995 Mekong Agreement Article 5 stipulates that only notification is required for both intra-basin uses and inter-basin diversions for tributaries, whereas in the mainstream such uses and diversions may be subject to prior consultation, depending on the season, that is, whether it is wet or dry. This is an important difference to note compared with the Xayaburi dam's intra-basin use on the mainstream during wet and dry seasons, which requires prior consultation (Article 5 *Mekong Agreement* 1995). This requirement for the Xayaburi dam created the space for all parties, including NGOs, to provide their opinions on the dam.

On the mainstream of the Mekong River, Petrovietnam, a private firm in Vietnam, has a plan to develop the Luang Prabang dam, located upstream of the Xayaburi dam in Laos (ICEM 2010b: 30). While it was not possible to gain the government's view on this proposed hydropower dam during the fieldwork, some interviewees who are former government officials expressed their opinions that this dam was not a major concern for the government of Vietnam, and there is a high possibility that this dam plan would be cancelled (V4 2012; V6 2012; Rotha and Vannarith 2008). This development status of the Luang Prabang dam appears to be one of the factors allowing the Vietnamese government to take a strong position against the Xayaburi dam, discussed in more detail in Chapter 7.

Conclusion

This chapter provides a discussion of the physical and material conditions associated with the development of hydropower dams on the Mekong River, and particularly of the Xayaburi dam. Following the typology of goods suggested for the IAD framework introduced in Table 2.4, the Mekong River has the characteristics of a common-pool resource, where both subtractability of use and difficulty of excluding potential beneficiaries are high (Ostrom 2005: 24). The nature of this resource makes it difficult to manage competing interests among potential users, particularly when the resource crosses borders, and managing and mitigating impacts created by one's actions requires involvement of a diverse range of actors.

The review of historical perspectives of the governance of the Mekong River and evolving interests over development of hydropower dams on the

Mekong River provides important insights into understanding the contexts of current hydropower dam development plans, and the interests of different actors who potentially shape the strategies of NGO coalitions within the LMB. Three important contexts emerge through this review. The first is the resource context of the Mekong River. Although development of mainstream hydropower dams has been on the development agenda within regional and international governments since the 1950s, the mainstream of the LMB has been physically untouched until the construction of the Xayaburi hydropower dam commenced in 2012 (Ministry of Energy and Mines 2012). The riparian countries faced decades of civil wars during the second half of the twentieth century, which ironically resulted in preserving the free-flowing status of the LMB mainstream. It was due to this relatively pristine status of the Mekong River that attention by the international community focused on the construction of the Xayaburi dam, including many NGOs and civil society actors.

The second context is associated with the positions of key actors. International organizations and donors, particularly bilateral donors from the Western governments as well as development banks such as the World Bank and the ADB, historically have promoted hydropower development on the mainstream of the Mekong River. During the decades when hydropower dam development plans on the mainstream were stalled due to the civil wars, the positions of these international donors gradually shifted to a cautious stance about downstream impacts (MRC 2011a, 2011c). These new positions by international donors signalled a positive shift for NGOs in the region that are opposed to, or concerned about, hydropower dam development on the Mekong River. This is a particularly important change, especially in relation to the circumstances that NGOs faced during their advocacy work in the past over the Mekong's tributary dams. As an example, the World Bank, which financed the controversial Pak Mun dam and Nam Theun II dam built on the tributaries of the Mekong, welcomed the SEA of the Mekong's mainstream dams, and confirmed that it will not finance any mainstream hydropower projects (World Bank 2010).

Finally, the analysis of the biophysical and material conditions in Cambodia and Vietnam clarifies that both countries' governments face competing interests over hydropower development and their impacts on natural resource use. These biophysical material conditions affect the positions of the governments in discussions of the Xayaburi hydropower dam, which is discussed further in Chapter 7.

References

ADB. 2012. *Greater Mekong Subregion Power Trade and Interconnection: 2 Decades of Cooperation.* Manila: Asian Development Bank. www.adb.org/sites/default/files/pub/2012/gms-power-trade-interconnection.pdf. Accessed 2 April 2013.

Agreement on the Cooperation for the Sustainable Development of the Mekong River Basin. 1995.

Allen, Joe. 2008. *Vietnam: The (Last) War the U.S. Lost.* Chicago, IL: Haymarket.

Bakker, Karen. 1999. The politics of hydropower: developing the Mekong. *Political Geography* 18: 209–232.

Baran, Eric, Teemu Jantunen, and Chong Chiew Kieok. 2007. *Values of Inland Fisheries in the Mekong River Basin.* Phnom Penh: WorldFish Center. www.worldfishcenter.org/resource_centre/WF_895.pdf. Accessed 20 April 2014.

C7. 2012. Personal interview, 27 July 2012.

C27. 2012. Personal interview, 9 August 2012.

Foran, Tira. 2006. Rivers of contention: Pak Mun Dam, electricity planning, and state–society relations in Thailand 1932–2004. PhD thesis, Division of Geography, School of Geosciences, University of Sydney, Australia.

Foster, Alice. 2011. Environmentalist slam Lao Dam impact studies. *Cambodia Daily* 15 March.

Geospatial Information Authority of Japan. undated. *Map of Geospatial Information Authority.* http://maps.gsi.go.jp/#5/19.290406/104.326172. Accessed 16 February 2015.

Graf, William L. 2006. Downstream hydrologic and geomorphic effects of large dams on American rivers. *Geomorphology* 79: 336–360.

Grimsditch, Mark. 2012. *3S Rivers Under Threat.* International Rivers. www.internationalrivers.org/files/attached-files/3s_rivers_english.pdf. Accessed 7 April 2013.

Hirsch, Philip. 2010. The changing political dynamics of dam building on the Mekong. *Water Alternatives* 3 (2): 312–323.

Hortle, K.G. 2007. *Consumption and the Yield of Fish and other Aquatic Animals from the Lower Mekong Basin.* MRC Technical Paper No. 16. Vientiane: Mekong River Commission. www.mrcmekong.org/assets/Publications/technical/tech-No16-consumption-n-yield-of-fish.pdf. Accessed 15 September 2012.

ICEM. 2010a. *MRC Strategic Environmental Assessment (SEA) of Hydropower on the Mekong Mainstream: Summary of the Final Report.* Hanoi: International Centre for Environmental Management. www.mrcmekong.org/assets/Publications/Consultations/SEA-Hydropower/SEA-FR-summary-13oct.pdf. Accessed 7 June 2015.

—— 2010b. *MRC Strategic Environmental Assessment (SEA) of the Hydropower on the Mekong Mainstream.* Mekong River Commission. Hanoi, Vietnam. www.mrcmekong.org/assets/Publications/Consultations/SEA-Hydropower/SEA-Main-Final-Report.pdf. Accessed 3 May 2014.

International Rivers. 2002. *Damming the Sesan River: Impacts in Cambodia and Vietnam.* www.internationalrivers.org/files/attached-files/04.sesan.pdf. Accessed 7 April 2013.

Jacobs, Jeffrey W. 1995. Mekong Committee history and lessons for river basin development. *Geographical Journal* 161 (2): 135–148.

—— 2002. The Mekong River Commission: transboundary water resources planning and regional security. *Geographical Journal* 168 (4): 354–364.

Joffre, Olivier, Mam Kosal, Yumiko Kura, Pich Sereywath, and Nao Thuok. 2012. *Community Fish Refuges in Cambodia – Lessons Learned.* Phnom Penh: Worldfish Center. www.worldfishcenter.org/resource_centre/WF_3147.pdf. Accessed 3 May 2014.

Joint Declaration of Principles for Utilization of the Waters of the Lower Mekong Basin, signed by the representatives of the governments of Cambodia, Laos, Thailand and Vietnam to the Committee for Coordination of Investigations of the Lower Mekong Basin, at Vientiane on 31 January 1975.

Joo, L. 2006. Mekong-map. http://commons.wikimedia.org/wiki/File: Mekong-map.jpg. Accessed 16 February 2015.

Keay, John. 2005. The Mekong Exploration Commission, 1866–68: Anglo–French rivalry in South East Asia. *Asian Affairs* 36 (3): 289–312.

Kingdom of Cambodia. 2011. *Mekong River Commission Procedures for Notification, Prior Consultation and Agreement: Form/Format for Reply to Prior Consultation.* Cambodia National Mekong Committee. www.mrcmekong.org/assets/Consultations/2010-Xayaburi/Cambodia-Reply-Form.pdf. Accessed 25 November 2011.

Kummu, Matti, and Juha Sarkkula. 2008. Impact of the Mekong River flow alteration on the Tonle Sap flood pulse. *Ambio* 37 (3).

Mekong Secretariat. 1994. *Annual Report 1994.* www.mrcmekong.org/assets/Publications/governance/Annual-Report-1994.pdf. Accessed 3 May 2014.

Mekong Watch. 2002. *Pak Mun Dam.* www.mekongwatch.org/english/country/thailand/pakmun.html. Accessed 24 August 2013.

Middleton, Carl, Jelson Garcia, and Tira Foran. 2009. Old and new hydropower players in the Mekong region: agendas and strategies. In *Contested Waterscapes in the Mekong Region: Hydropower, Livelihoods and Governance,* edited by F. Molle, T. Foran, and M. Käkönen. London/Sterling, VA: Earthscan.

Ministry of Energy and Mines. 2012. Statement of the Government of the Lao PDR on Xayaburi Hydropower Project. www.poweringprogress.org/index.php?option=com_content&view=article&id=947:19th-november-2012-statement-of-the-government-of-lao-pdr-on-xayaburi-hydropower-project&catid=85:press-releases&Itemid=50. Accessed 28 March 2013.

Ministry of Planning and Investment. 2011. *The Seventh Five-year National Socio-Economic Development Plan (2011–2015).* www.drrgateway.net/sites/default/files/LAO%20 5Year%20Dev%20pLan%2020112015.pdf. Accessed 29 March 2013.

Moore, Deborah, John Dore, and Dipak Gyawali. 2010. The World Commission on Dams + 10: Rivisiting the Large Dam Controversy. *Water Alternatives* Volume 2 (Issue 2): 3–13.

MRC. 2003. Biodiversity and Fisheries in the Mekong River Basin. Mekong Development Series No. 2. Mekong River Commission. www.mrcmekong.org/assets/Publications/report-management-develop/Mek-Dev-No2-Mek-River-Biodiversityfiisheries-in.pdf. Accessed 18 January 2014.

—— 2006a. Annual Report 2006. www.mrcmekong.org/assets/Publications/governance/annual-report-2006.pdf. Accessed 3 May 2014.

—— 2006b. Minutes of the twenty-fourth meeting of the Joint Committee (29–30 August 2006, Vientiane, the Lao PDR).

—— 2008. Annual Report 2008. www.mrcmekong.org/assets/Publications/governance/Annual-Report-2008.pdf. Accessed 3 May 2014.

—— 2010a. Assessment of Basin-wide Development Scenarios. Main Report. www.mrcmekong.org/programmes/bdp/Tech-Notes/Assessment-of-Basin-wide-Dev-Scenarios-Main-ReportPt-1-2-101103.pdf. Accessed 15 April 2011.

—— 2010b. State of the Basin Report 2010: Summary. www.mrcmekong.org/assets/Publications/basin-reports/MRC-SOB-Summary-reportEnglish.pdf. Accessed 11 September 2012.

—— 2011a. Joint Development Partner Statement: Donor Consultative Group – 18th MRC Council Meeting, Joint Meeting with the MRC Donor Consultative Group. www.mrcmekong.org/news-and-events/speeches/joint-development-partner-statement-donor-consultative-group-18th-mrc-council-meeting-joint-meeting-with-the-mrc-donor-consultative-group/. Accessed 15 December 2011.

—— 2011b. Prior Consultation Project Review Report: Volume 2 Stakeholder Consultations related to the proposed Xayaburi dam project. www.mrcmekong.org/

pnpca/2011-03-31-Report-on-Stakeholder-Cons-on-Xayaburi.pdf. Accessed 11 April 2011.

—— 2011c. Report: Informal Donor Meeting. 23–24 June 2011. Phnom Penh, Cambodia. www.mrcmekong.org/assets/Publications/governance/Minutes-of-IDM2011-final. pdf. Accessed 31 October 2011.

—— 2012. Minutes of the Eighteenth Meeting of the MRC Council (8 December 2011, Siem Reap, Cambodia). www.mrcmekong.org/assets/Publications/governance/ Minutes-of-the-18th-Council.pdf. Accessed 29 September 2012.

—— 2013a. Joint Development Partner Statement: 19th MRC Council Meeting, 17 January 2013. www.mrcmekong.org/news-and-events/speeches/joint-development-partner-statement-19th-mrc-council-meeting-17-january-2013/. Accessed 4 May 2013.

—— 2013b. Statement by H.E. Dr. Nguyen Thai Lai – 19th Meeting of the MRC Council. www.mrcmekong.org/news-and-events/speeches/statement-by-h-e-dr-nguyen-thai-lai-19th-meeting-of-the-mrc-council/. Accessed 15 July 2013.

—— 2013c. Statement by H.E. Mr. Sin Niny – 19th Meeting of the MRC Council. www. mrcmekong.org/news-and-events/speeches/statement-by-h-e-mr-sin-niny-19th-meeting-of-the-mrc-council/. Accessed 15th July 2013.

—— undated. The Mekong Basin: Physiology. www.mrcmekong.org/the-mekong-basin/ physiography/. Accessed 2 April 2013.

MRC Secretariat. 2011. Proposed Xayaburi Dam Project-Mekong River: Prior Consultation Project Review Report. www.mrcmekong.org/assets/Publications/Reports/PC-Proj-Review-Report-Xaiyaburi-24-3-11.pdf. Accessed 25 October 2011.

MWBP. 2005. About the programme. Mekong Wetlands Biodiversity Conservation and Sustainable Use Programme. www.mekongwetlands.org/Programme/mekong.htm. Accessed 18 January 2014.

Nam Theun 2 Power Company. undated. Welcome to Nam Theun 2 Power Company (NTPC). www.namtheun2.com/. Accessed 19 March 2013.

Nippanon, J., R. Schouten, P. Sripapatrprasite, C. Vaddhanaphuti, C. Vidthayanon, W. Wirojanagud, and E. Watana. 2000. Pak Mun dam, Mekong river basin, Thailand. A WCD Case Study prepared as an input to the World Commission on Dams. World Commission on Dams. Cape Town. www.centre-cired.fr/IMG/pdf/F8_PakMunDam. pdf. Accessed 11 January 2012.

Orr, Stuart, Jamie Pittock, Ashok Chapagain, and David Dumaresq. 2012. Dams on the Mekong River: Lost fish protein and the implications for land and water resources. *Global Environmental Change* 22 (4): 925–932.

Ostrom, Elinor. 2005. *Understanding Institutional Diversity*. Princeton: Princeton University Press.

Quang, Nguyen Hhan. 2002. Vietnam and the sustainable development of the Mekong river basin. *Water Science & Technology* 45 (11): 261–266.

Rotha, Chan, and Chheang Vannarith. 2008. Cultural challenges to the decentralization process in Cambodia. *Ritsumeikan Journal of Asia Pacific Studies* 24: 1–16.

Sadalmelik. 2007. Topographic map of Vietnam. Created with GMT from publicly released GLOBE data. http://en.wikipedia.org/wiki/File: Vietnam_Topography.png. Accessed 2 May 2014.

Save the Mekong. 2011. Subject: Request for MRC Council to halt the current PNPCA process on the Xayaburi Dam, to endorse the MRC's SEA report's findings, and commit to evaluate all options for meeting the Mekong region's water and energy needs through a credible and objective public process. A letter to H.E. Mr Lim Kean Hor: Minister of Water Resources and Meteorology et al., 25 January.

Socialist Republic of Viet Nam. 2011. Mekong River Commission Procedures for Notification, Prior Consultation and Agreement Form for Reply to Prior Consultation. The Viet Nam National Mekong Committee. www.mrcmekong.org/assets/ Consultations/2010-Xayaburi/Viet-Nam-Reply-Form.pdf. Accessed 25 October 2011.

Stuart-Fox, Martin. 1997. *A History of Laos*. Cambridge: Cambridge University Press.

TEAM Consulting Engineering and Management Co. Ltd. 2008. *Feasibility Study Xayaburi Hydroelectric Power Project Lao, PDR*. Final Report (Main Report). CH. Karnchang Public Company Limited. www.mrcmekong.org/assets/Consultations/2010-Xayaburi/ xayaboury-dam-feasibility-study.pdf. Accessed 25 October 2011.

—— 2010. *Environmental Impact Assessment. Xayaburi Hydroelectric Power Project. The Lao PDR*. CH. Karnchang Public Company Limited. www.mrcmekong.org/assets/ Consultations/2010-Xayaburi/Xayaburi-EIA-August-2010.pdf. Accessed 25 October 2011.

UNESCO. 2013. *Town of Luang Prabang*. http://whc.unesco.org/en/list/479. Accessed 28 March 2013.

V4. 2012. Personal interview, 11 July 2012.

V6. 2012. Personal interview, 12 July 2012.

V18. 2012. Personal interview, 14 August 2012.

Vickery, Michael. 1984. *Cambodia 1975–1982*. Sydney: South End Press.

WCD. 2000. *Dams and Development: A New Framework for Decision-Making*. London: Earthscan.

World Bank. 2010. *World Bank Group Welcomes Strategic Environmental Assessment of Mekong Mainstream Dams*. http://web.worldbank.org/WBSITE/EXTERNAL/COUNTRIES/ EASTASIAPACIFICEXT/CAMBODIAEXTN/0,,contentMDK:22740418~menuPK :293875~pagePK:2865066~piPK:2865079~theSitePK:293856,00.html. Accessed 18 January 2014.

WWF. 2013. *Greater Mekong*. http://worldwildlife.org/places/greater-mekong. Accessed 14 July 2013.

Wyatt, Andrew B., and Ian G. Baird. 2007. Transboundary impact assessment in the Sesan River Basin: the case of the Yali Falls dam. *International Journal of Water Resources Development* 23 (3): 427–442.

5   Actors

Introduction

During the Procedures for Notification, Prior Consultation and Agreement (PNPCA) of the Xayaburi hydropower dam, a wide range of actors played different roles. This chapter introduces key actors associated with the Xayaburi dam and the advocacy strategies of the case-study NGO coalitions. The chapter is divided into three parts. The first section introduces actors working across the Mekong region, particularly within the Lower Mekong Basin. The chapter then introduces actors within Cambodia, followed by actors in Vietnam.

Regional actors

Mekong River Commission

As discussed in Chapter 4, the MRC was established as an institutional framework for cooperation between the four states that are parties to the 1995 Mekong Agreement: Cambodia, Thailand, Laos, and Vietnam (Article 11 *Mekong Agreement* 1995). The MRC consists of three bodies: the Council, the Joint Committee (JC), and the Secretariat (Article 12 *Mekong Agreement* 1995). The Council consists of representatives of each riparian state from Ministerial and Cabinet levels, who have the power to make decisions on behalf of the government (Article 15 *Mekong Agreement* 1995). The JC consists of representatives who are at least Head of Department level (Article 21 *Mekong Agreement* 1995), and assumes responsibility for implementing policies and decisions made by the Council (Article 24 *Mekong Agreement* 1995). Each member country establishes a National Mekong Committee (NMC), which is responsible for matters associated with the Mekong River. The NMCs of each member state were the main actors in the riparian state governments' dealings with the Xayaburi hydropower dam.

The MRC Secretariat provides technical and administrative services to the Council and the JC (Article 28 *Mekong Agreement* 1995). The Secretariat is directed by the CEO selected by the JC and appointed by the Council (Article 31 *Mekong Agreement* 1995). As of 2010, the Secretariat consisted of 153 staff including nine international professionals (Taylor and MRC Secretariat 2011:

63). Despite its expertise and a large number of staff, the Secretariat does not have decision-making powers within the scheme of the MRC. Decisions at the MRC are made through unanimous vote by member states (Articles 20 & 27 *Mekong Agreement* 1995).

As a transboundary water-governance mechanism based on international agreements, all the decisions taken by the MRC Council are, in principle, decided by the consensus of all member countries (Articles 20 & 27 *Mekong Agreement* 1995). It is important to stress that the MRC Secretariat is not a decision-making body. The MRC Secretariat has often been the target of criticism by NGOs and civil society actors due to its perceived inability to make decisions. It is the MRC Council that is a forum for decision-making among states, not the Secretariat (R10 2011). As one interviewee from the NGO sector commented:

> [In our] 15 years of strong criticism of the MRC, there is a misunderstanding about what the MRC is. We should really be criticising our own government. The (staff of) MRC Secretariat, they are just advisors.
>
> (R8 2012)

Donors to the Mekong River Commission

The activities of the MRC are supported by financial and technical contributions from member and non-member countries, as well as other development partners. As of 2012, the external donors to the MRC included the governments of Australia, Belgium, Denmark, the European Union (EU), Finland, France, Germany, Japan, Luxembourg, the Netherlands, New Zealand, Sweden, Switzerland, the United States, and the World Bank (MRC 2013a). Germany, Finland, the Netherlands, and Australia are the top donors based on their commitment during 2011–15, which constituted three-quarters of all donor funds (MRC 2011: 52). In addition to donors, the MRC has Memoranda of Understanding (MoUs) with several organizations for technical cooperation. These MoUs have been concluded primarily with international and regional organizations such as the Asian Development Bank (ADB), Association of Southeast Asian Nations (ASEAN), EU, International Union for Conservation of Nature (IUCN), United Nations Development Programme (UNDP), and United Nations Economic and Social Commission for Asia and the Pacific (UNESCAP) (MRC 2011, undated-a). These partners are invited to observe the MRC Council meetings and JC meetings (MRC undated-a). The MRC and development partners also conduct informal donor meetings with the MRC member countries annually, creating the opportunity to raise any issues of concern. These meetings also provide opportunities for coordinating the efforts of partners in support of the MRC (MRC 2011).

The World Wide Fund for Nature (WWF) is the only NGO among the MRC's development partners, positioning itself differently from other NGOs within the region. The WWF signed an MoU with the MRC in 2002, and since then has conducted various collaborative projects including promoting sustainability in hydropower development (MRC 2006: 46–47). The IUCN is another partner organization to the MRC with a focus on environment and conservation. The IUCN consists of both governmental and nongovernmental member organizations (IUCN 2013; MRC 2006). It is important to note that WWF was not invited to the MRC Council meeting that took place in January 2013, shortly after Laos officially launched the construction of the Xayaburi dam. Donors to the MRC expressed their concern about this sudden un-invitation to one of the MRC partners, and expressed their concerns through the Joint Development Partner Statement presented at the meeting (MRC 2013).

Save the Mekong Coalition

The Save the Mekong (STM) Coalition is a regional network of NGOs, community groups, academics, and ordinary citizens within the Mekong region and internationally, who share concerns regarding the future of the Mekong River (Save the Mekong 2009). The coalition was officially launched in 2009 at the time when Laos' plan for constructing the Don Sahong dam, another dam planned on the Mekong's mainstream, was progressing. However, the personal and organizational networks among activists existed in the region prior to the official launch of the STM Coalition (R10 2011; R3 2012). The members of the STM Coalition have been campaigning against hydropower dams on the mainstream of the Mekong River since before the official launch of the coalition (Save the Mekong 2009; R10 2011). While the STM Coalition does not have an official coordinator, International Rivers and Towards Ecological Recovery and Regional Alliance (TERRA) act as informal coordinators of the network (V2 2012; C5 2012; V11 2012). International Rivers is an international NGO headquartered in the USA, campaigning against large-scale development on rivers around the globe; TERRA is a Thai-based environmental NGO which has been engaged in anti-dam campaigns in Thailand since the 1980s (International Rivers undated; TERRA undated). Both NGOs have offices in Bangkok and have been playing important roles in fostering the regional network of NGOs around the Mekong region. The STM Coalition is one of the most active and vocal NGO networks in the Mekong region over the issue of the Xayaburi dam, and the two case-study NGO coalitions of this book, the Rivers Coalition in Cambodia (RCC) and the Vietnam Rivers Network (VRN), are both members of this regional coalition (Save the Mekong undated).

Developers of the Xayaburi dam

The development of the Xayaburi dam is mostly driven by actors in Thailand. The installed capacity of the dam will be 1280 MW, and most of the energy produced from the Xayaburi dam will be exported to Thailand (TEAM

Consulting Engineering and Management Co. Ltd. 2008: 1-2, 1-4). The Electricity Generating Authority of Thailand (EGAT), which is a state enterprise under the Thai Ministry of Energy (EGAT 2009), promised to purchase 95 per cent of the electricity produced by the dam, and signed the power purchase agreement with the Xayaburi Power Company Limited in October 2011 (International Rivers 2013a, 2013b).

The Xayaburi Power Company is the main developer of the Xayaburi hydropower dam, it was established in 2010 by CH. Karnchang Public Company Limited (CH. Karnchang), which is the second largest construction Company in Thailand (TEAM Consulting Engineering and Management Co. Ltd. 2008: 1-1; Middleton 2012: 309; CH. Karnchang 2011: 143). CH. Karnchang holds a majority of the ownership of the Xayaburi Power Company (Middleton 2012: 304; International Rivers 2013b; CH. Karnchang 2011: 55, 143). CH. Karnchang is also the company responsible for the construction of the Xayaburi dam (CH. Karnchang 2012).

CH. Karnchang signed an MoU with the Government of Laos to conduct a feasibility study for the Xayaburi dam in 2007 (TEAM Consulting Engineering and Management Co. Ltd. 2008: 1-1). This was used as a basis for signing a project development agreement between the Government of Laos and CH. Karnchang in November 2008 (International Rivers 2011). The Government of Laos granted the Xayaburi Power Company a 29-year concession to design, develop, construct, and operate the Xayaburi dam, commencing from the commercial operation date (CH. Karnchang 2011: 143). The development takes place through a build–operate–transfer (BOT) arrangement, which is a scheme allowing the private sector to build and operate an infrastructure asset and eventually transfer the ownership to the public sector (TEAM Consulting Engineering and Management Co. Ltd. 2008). The finance for the Xayaburi dam comes from Thai banks including Kasikorn Bank, Bangkok Bank, Krung Thai Bank, and Siam Commercial Bank (Middleton 2012: 304; International Rivers 2011). The Kasikorn Bank is a state-owned bank, and its financing of the Xayaburi dam was challenged by Thai civil society, as discussed later in Chapter 7.

Actors in Cambodia

Rivers Coalition in Cambodia

The study in Cambodia focuses on the RCC, a network of NGOs concerned with environmental and human rights issues related to hydropower dams (Rivers Coalition in Cambodia undated: 1). The network was established in 2003 and was originally called the Se San Working Group. This network started as a small number of NGOs working together on the hydropower dam issues in the Se San River, one of the tributaries of the Mekong River (Rivers Coalition in Cambodia undated). The initial focus of the network was on addressing livelihoods, and the health impacts on Cambodian communities downstream of the dam built on the Vietnamese side of the border. The original members included the NGO Forum on Cambodia (hereafter NGO Forum), the 3S Rivers Protection Network (3SPN), and the Culture and Environment Preservation Association (CEPA) (Rivers Coalition in Cambodia undated: 1; C3 2012; C11 2012). More recently, the network expanded its scope to all the dams in Cambodia and changed its name to Rivers Coalition in Cambodia, and in turn expanded its membership (NGO Forum undated; C3 2012). As of December 2011 the network had 28 Cambodian NGOs as its members (NGO Forum 2011). In addition, the RCC has 14 international partners (Rivers Coalition in Cambodia 2012; NGO Forum 2011). While initial members were primarily advocacy-oriented NGOs, the newer members of the RCC include NGOs with a focus on improvement of rural livelihoods, conservation, and human rights (C28 2012; C11 2012; NGO Forum 2011). Some of these NGOs are located in the capital Phnom Penh, others in the provinces (NGO Forum 2011). This shift in membership composition was an important influence on the way the RCC operated during its Xayaburi advocacy work, as discussed in later chapters.

The RCC is coordinated by the Hydropower & Community Rights Project, one of the projects within the Environment Programme of the NGO Forum (NGO Forum undated). The NGO Forum is an organization of Cambodian NGOs. As of 2013, the Forum had 89 members, all NGOs operating in Cambodia. The NGO Forum serves primarily as a forum for sharing information among NGOs, and takes joint actions through various programmes (NGO Forum 2013). The NGO Forum has been engaged in supporting communities affected by hydropower developments since the late 1990s (NGO Forum undated). The coordinator of the NGO Forum's hydropower project also coordinates the RCC (C5 2012). The NGO Forum provides some funding for the RCC's activities, in addition to RCC specific donors such as Oxfam Australia, the EU, and Internationaal Maatschappelijk Verantwoord Ondernemen [Interchurch Organization for Development Cooperation, ICCO] (C3 2012; C5 2012). When RCC members travel to Phnom Penh for their regular meetings, the travel costs for members based in the provinces are covered by the funding for the RCC administered by the NGO Forum (C3 2012). While each member NGO is expected to contribute to the network, the current institutional set-up of the RCC places much of the workload and financial responsibility on the NGO Forum.

Cambodian government

The Cambodian Constitution addresses the separation of legislative, judiciary, and executive powers (Article 51 *Constitution of Cambodia* 1993). The National Assembly and the Senate hold legislative power (Articles 90, 99 *Constitution of Cambodia* 1993). The National Assembly has the function of approving various policy measures proposed by the executive, including the budget, state plans, administration, and international treaties. It can also pass a vote of confidence on the Royal Government (Article 90 *Constitution of Cambodia* 1993). The executive power lies with the Royal Government, which consists of the Council of Ministers. It is led by the Prime Minister, with members including Deputy Prime Ministers, State Ministers, Ministers, and State Secretaries (Article 118 *Constitution of Cambodia* 1993). Cambodia is a multi-party state, and the head of the leading political party assumes the position of Prime Minister. The current Prime Minister is Hun Sen, head of the Cambodian People's Party (CPP), who has been leading the country since 1985 (Vandenbrink 2013; BBC News Asia 2013). The judiciary power is independent, and lies with the Supreme Court and other courts (Article 128 *Constitution of Cambodia* 1993). However, in reality separation of three powers – judiciary, executive, and legislative – is blurred. This remains a key challenge for implementing the rule of law and protecting human rights in Cambodia (paragraph 41 United Nations Human Rights Council 2010: 11).

Administratively, the country is divided into provinces and municipalities. Provinces are divided into districts (*srok*) then subdivided into communes (*khum*); municipalities are divided into sections (*khans*) then subdivided into quarters (*sangkats*) (Article 145 *Constitution of Cambodia* 1993; Obendorf 2004: 10). The communes and *sangkats* are the smallest administrative units in the country, and the most local levels of democracy. The commune and *sangkats* councils have both legislative and executive authorities within their local jurisdiction, allowing the councils to adopt resolutions that apply to the governance of the locality (Article 48 *Law on Commune Management* 2001). The councils are also responsible for producing commune/*sangkat* development plans and budgets, and have the right to impose local fiscal taxes (Article 51 *Law on Commune Management* 2001). The members of commune and *sangkat* councils are elected, with elections taking place every five years (Article 3 *Amended Law on Elections of Commune Councils* 2006).

Cambodian National Mekong Committee

The CNMC represents Cambodia in the MRC. It is currently chaired by the Ministry of Water Resources and Meteorology (MOWRAM), which has the mandate for river basin management in Cambodia, including international waters (Articles 10, 34 *Law on Water Resources Management* 2007). The CNMC's mission is to domestically and internationally coordinate the protection, conservation, and development of resources within the Mekong River Basin (Cambodian National Mekong Committee undated). The CNMC operates directly under the Royal Government of Cambodia, and is comprised of ten water-related ministries (Cambodian National Mekong Committee undated; Hirsch *et al.* 2006: 41). The Minister of MOWRAM is the chairperson of the CNMC, and is a member of the MRC Council (MRC undated-b). The Secretary General of the CNMC represents Cambodia as a member of the MRC Joint Committee (MRC undated-b).

Actors in Vietnam

Vietnam Rivers Network

The VRN is the study coalition in Vietnam for this book. It is an open forum comprised of members concerned with the protection and sustainable development of rivers in Vietnam (Vietnam Rivers Network 2009: 4). The network was established in 2005 and was initially hosted by the Institute of Ecological Economy (Eco-Eco) (Vietnam Rivers Network 2009: 4). The original funding came from the Siemempuu Foundation in Finland, and the founder of the network was supported by International Rivers (V10 2012). In 2007, the network changed its name to the Vietnam Rivers Network, and a new Vietnamese NGO, the Centre for Water Resources Conservation and Development (WARECOD), was established with the main purpose of hosting the VRN (TERRA 2008: 32; V10 2012).

Membership of the VRN comprises NGOs, researchers, academics, government officials, local communities, and individuals (Vietnam Rivers Network 2009: 4). As of November 2012, the network had approximately 300 members (Vietnam Rivers Network 2012). In contrast with the RCC in Cambodia, where the network consists solely of NGOs, the majority of VRN members are individuals who are interested in and concerned about issues related to rivers in Vietnam (Vietnam Rivers Network 2012). As of 2012, there are eight organizational members that are considered key members of the VRN (Vietnam Rivers Network undated-b; V23 2012).

Some of the organizational members focus their work on issues of water and energy, while others have a focus on natural resources management and sustainable livelihoods (Vietnam Rivers Network undated-b). Many of the members are Vietnamese NGOs who are relatively newly established, most of them first established in the mid-2000s. One of the key organizational members is a government entity. The Centre for Biodiversity and Development (CBD), which coordinates the VRN's activities in the southern part of Vietnam, was established under the Institute of Tropical Biology as a non-profit, government-funded institution (V12 2012). The network also includes many individual members who are government officials. The mixed nature of the network, including both government and non-government actors, influences its characteristics.

The network is organized into three regional networks: northern, central, and southern Vietnam. Each network is coordinated by one of the key member NGOs: the northern network is coordinated by WARECOD, the central network by CSRD, and the southern network by CBD. The VRN is designed to rotate its overall coordinating role among these three regional organizations. Based on this mechanism, the overall coordination of the VRN was moved to the CSRD in 2012 (Vietnam Rivers Network 2013). Many of the national networks are typically based in Hanoi, and the VRN's initiative to rotate the coordination is a rather unique approach in Vietnam, which, according to one interviewee, supports strengthening central and southern regional networks

(V11 2012). According to another interviewee, the rotation also prevents the VRN from representing concerns limited to one region within the country, or from relying only on support from a few organizations (V10 2012).

The VRN's specific issue-based activities are conducted through thematic and regional groups within the network. As of February 2013, there were 11 groups within the network (Vietnam Rivers Network undated-a). The Mekong task force was one of the groups within the VRN, which worked primarily on the Xayaburi hydropower dam. According to one interviewee, this was one of the most active groups within the network (V5 2012). The Mekong task force included the VRN coordinator, a staff member of WARECOD, a former high-ranking official of the Vietnam National Mekong Committee (VNMC), and scientists specializing in the ecosystem of the Mekong Delta (V10 2012).

Vietnamese government

Vietnam is a socialist country with a highly centralized government system. The legislative responsibility lies solely with the National Assembly, and its members are elected through general elections which are intended to provide the mechanism for interest articulation (Article 83 *Constitution of Vietnam* 2001). The National Assembly adopts and amends laws, decides on national socio-economic plans, and decides on financial matters, including budget and tax (Article 83 *Constitution of Vietnam* 2001). The National Assembly also elects or suspends the main political figures in the executive and judiciary, including the Prime Minister, other Ministers, the President, the Chairman, and the members of the standing committee of the National Assembly (Article 84 *Constitution of Vietnam* 2001). The standing committee of the National Assembly organizes and chairs the National Assembly and supervises the activities of the People's Council at province and local levels (*Constitution of Vietnam* 2001).

The executive function is conducted by the government, which consists of the Prime Minister, Deputy Prime Ministers, Ministers and heads of ministerial-level agencies (Article 3 *Law on Organization of the Government. No. 32/2001/ QH10 of 25 December 2001* 2001). The government is responsible for submitting draft laws and policies to the National Assembly, and it implements laws and policies. It also leads the work of the ministries, including the sub-national levels of bureaucracy such as at the province, district, and commune levels. The judiciary is composed of the Supreme People's Court, Provincial People's Court, District People's Court, Military Tribunals, and other Tribunals (Article 127 *Constitution of Vietnam* 2001).

The country is divided into provinces, then further into districts, provincial cities, and municipalities. These sub-national administrative units are further divided into wards and communes (Article 118 *Constitution of Vietnam* 2001). Communes are composed of several villages and are also the lowest level of administrative unit. People's Councils are the authorities that are elected by a local population, and are responsible for socio-economic development plans and budgets for their localities, ensuring local security, and conducting any other tasks assigned by higher authorities of the state (Article 120 *Constitution of Vietnam* 2001). People's Committees are the executive agencies of the respective People's Councils, and are elected by the People's Councils (Article 123 *Constitution of Vietnam* 2001).

Vietnam is a one-party state led by the Communist Party of Vietnam (CPV). While the Constitution states that people are the 'master' of the country (Article 2 *Constitution of Vietnam* 2001), the leading role of the CPV is also clearly stated in the Constitution (Article 4 *Constitution of Vietnam* 2001), allowing the formal system to be structured in a way that gives the CPV the power to lead the country. The party's political bureau, which consists of high-ranking officials from the CPV, plays key roles in the governance of the country. Scholars claim that most laws and policies discussed at the National Assembly are drafted by the party's political bureau and government offices behind closed doors, and public deliberations are rare (Cima 1987a; Kerkvliet 2001). Politicians are elected through elections at different levels, including at the commune, province, and national levels (Embassy of Socialist Republic of Vietnam in the USA 2013). In most cases, candidates are limited to Communist Party members, giving limited choice for the electorate to vote for other candidates. While the Constitution allows that all Vietnamese citizens have a right to vote and stand for election, regardless of their status or affiliation with political organizations (Article 54 *Constitution of Vietnam* 2001), the most high-ranking government positions are occupied by members of the CPV (Cima 1987b). The CPV plays an important role in determining the policies of the government. Access to decision-makers at national and provincial levels is very restricted (Kerkvliet 2001: 246), which makes it particularly important to have a personal 'entry point' in working with the government in Vietnam. It will be shown later in this book that this relates to the strategy the VRN has taken in its advocacy approach (see Chapter 9).

Mass organizations

In addition to the formal administrative systems at local level, it is important to note the existence and role of mass organizations in Vietnam. Originally established by the Indochinese Communist Party as a way to include all sectors in the anti-colonial struggle (UNDP Vietnam 2006: 7), mass organizations are occupation-based associations, such as the women's union, farmers' union, trade union, and youth union, established as early as the 1930s under the umbrella of the Vietnam Fatherland Front (VFF) (Norlund *et al.* 2006). Mass organizations were structured within the state system since the mid-1950s, as Vietnam established a highly centralized system of government with three pillars: the party, the government and the mass organizations (Sinh 2014: 43).

The VFF is an umbrella organization of 29 mass organizations (Thayer 2008: 3; Norlund 2007: 11) and plays a role as the 'civil society arm' of the CPV. The Vietnamese Constitution defines the VFF as a politically allied organization and voluntary union of political organizations which represents individuals from different social strata, classes, ethnic groups, and religions (Article 9 *Constitution of Vietnam* 2001). The VFF is officially positioned to be able to provide inputs to the government. For example, the Constitution provides the VFF and its member organizations with the right to submit draft laws to the National Assembly (Article 87 *Constitution of Vietnam* 2001).

At the local level, the VFF plays an important role in local politics. A key role of the VFF is to conduct 'mass mobilization' in order to communicate government policies to people and to help them understand and implement the policies (UNDP Vietnam 2006: 7). Mobilization of peasants and workers aimed at supporting the state's policies is considered as local-level 'participation' in the Vietnamese political context (UNDP Vietnam 2006: 7). The VFF also plays a role in elections through organizing consultations to select and nominate candidates, supervise elections, and mobilize voters for the elections (Article 8 *Law on Vietnam Fatherland Front. No:14/1999/QH10 of 12 June 1999* 1999). In reality, these nominations are dominated by the CPV, and mostly produce candidates approved by the party leaders in the localities, or the party's Central Committee for candidates standing for higher offices (Kerkvliet 2001: 245).

Farmers' Union

The Farmers' Union is one of the mass organizations existing throughout Vietnam; it has approximately eight million members (Norlund *et al.* 2006: 33). It is a socio-political organization under the leadership of the CPV (Viet Nam Farmers' Union undated) and is characterized as a mass organization under the umbrella of the VFF (Norlund *et al.* 2006: 33).

Vietnam Union of Science and Technology Associations

VUSTA is a socio-political organization of Vietnamese intellectuals and scientists established in 1983 (Norlund *et al.* 2006: 25; VUSTA 2012a, b). VUSTA is a member of the VFF (VUSTA 2012b). Under the umbrella of VUSTA there are 73 member science and technology associations, 300 affiliated science and technology organizations, 197 newspapers, magazines, bulletins, and websites, and 60 provincial unions of science and technology associations (VUSTA undated: 9). WARECOD is one of the organizations affiliated with VUSTA.

Vietnam National Mekong Committee

In Vietnam, the primary responsibility for water resources lies with the Ministry of Natural Resources and Environment (MONRE), which chairs the VNMC. The main tasks of the VNMC include cooperation with other riparian countries to implement the 1995 Mekong Agreement, and the protection of Vietnamese interests concerning the use of the Mekong River, particularly on the mainstream (Article 2 *Decision No. 860-TTg* 1995). The VNMC is an inter-ministerial committee comprising MONRE, the Ministry of Foreign Affairs, the Ministry of Agriculture and Rural Development, and the Ministry of Planning and Investment (Article 3 *Decision No. 860-TTg* 1995). Other part-time members of the VNMC include the Ministry of Industry, the Ministry of Aquaculture, the Ministry of Science, Technology and Environment, the Ministry of Communications and Transport, the General Department of Meteorology and Hydrology, and representatives of People's Committees within the provinces associated with the Mekong River Basin (Article 3 *Decision No. 860-TTg* 1995).

Conclusion

The review of key actors in this chapter provides the background for the analysis of how actors interact with each other, and how they might influence the advocacy strategies of the case-study coalitions. The comparison of different actors within the context of Cambodia, Vietnam, and the Mekong region identify four key points, which are considered further in the analysis of advocacy strategies in Part III of this book.

First, among all the NGOs that exist within the Mekong region, the MRC has a formal partnership with only one, the WWF. The fact that other NGOs have no official relationship or status in association with the MRC places them as outsiders to the MRC.

Second, the RCC and the VRN are both coalitions of NGOs and civil society actors coordinated by an NGO. However, the difference in their memberships causes the two networks to have different characteristics. The RCC is primarily a coalition of NGOs, whereas the VRN is a coalition that includes both NGOs and individual members. The number of members is also different: the RCC has 28 members whereas the VRN has approximately 300. This difference in member characteristics is discussed further in the analysis of strategies in chapters 7–11.

Third, both coalitions were established through collaboration and support from international organizations. The RCC in particular was supported by multiple international organizations in its establishment phase. International Rivers was engaged in the establishment of both coalitions, and still continues to be an important partner for them (Vietnam Rivers Network 2013; Rivers Coalition in Cambodia undated; NGO Forum 2011).

Finally, differences and similarities in the government systems in Cambodia and Vietnam are highlighted in this chapter. Vietnam is a single-party state, where the role of the CPV is clearly stated in its Constitution. Cambodia, on the other hand, is a multi-party state, where the leading party is elected. However, in reality one party, the CPP, has dominated the country for the past two decades and, as commented by some scholars, Cambodia is becoming an authoritarian state (McCargo 2005; Un 2011).

In comparing the political contexts of Vietnam and Cambodia, it is important to note also that a force supported by Vietnam was the one that liberated Cambodia from the notorious Khmer Rouge regime which ended in 1979 (Vickery 1984). The current Cambodian Prime Minister, Hun Sen, was part of the Vietnamese troops and was positioned as deputy Prime Minister under the government installed by Vietnam in 1979 (BBC News Asia 2013; Nam *et al.* 2012). Due to this historical context, the Cambodian government has a close relationship with its Vietnamese counterpart and is considered to have a tendency to follow what the Vietnamese government says (C22 2012; C27 2012; McCargo 2005: 101). Some of the interviewees commented that this tendency was also observed in Cambodian government's position related to the Xayaburi dam (C22 2012; C27 2012).

References

1992 Constitution of the Socialist Republic of Vietnam: As Amended 25 December 2001 (transl. Allens Arthur Robinson). 2001. Vietnam.
Agreement on the Cooperation for the Sustainable Development of the Mekong River Basin. 1995.
Amended Law on Elections of Commune Councils. 2006. Cambodia.
BBC News Asia. 2013. Profile: Cambodia's Hun Sen. 26 July 2013. www.bbc.co.uk/news/world-asia-23257699 Accessed 11 May 2013.
C3. 2012. Personal interview, 25 July 2012.
C5. 2012. Personal interview, 25 July 2012.
C11. 2012. Personal interview, 28 July 2012.
C22. 2012. Personal interview, 7 August 2012.
C27. 2012. Personal interview, 9 August 2012.
C28. 2012. Personal interview, 10 August 2012.
Cambodian National Mekong Committee. undated. Welcome Address of H.E. Mr. Lim Kean Hor. www.cnmc.gov.kh/. Accessed 24 October 2012.
CH. Karnchang Public Company Limited. 2011. Building a better life: Annual Report 2010. CH. Karnchang Public Company Limited. www.ch-karnchang.co.th/2011/shareholder/ar_en10.pdf. Accessed 29 March 2013.
—— 2012. Notification of Contract Signing with Xayaburi Power Company Limited. www.ch-karnchang.co.th/news_activities_detail_en.php?nid=474. Accessed 19 October 2012.
Cima, Ronald J, ed. 1987a. *Vietnam: A Country Study.* Washington: GPO for the Library of Congress.
—— 1987b. The Vietnamese Communist Party. In *Vietnam: A Country Study.* Washington: GPO for the library of Congress.
The Constitution of the Kingdom of Cambodia. 1993. Cambodia.
Decision No. 860-TTg of 30 December 1995 of the Prime Minister on the Function, Tasks, Powers and Organization of the Apparatus of the Vietnam Mekong River Committee. 1995. Vietnam.
EGAT. 2009. EGAT: Electricity Generating Authority of Thailand. www.egat.co.th/en/. Accessed 29 March 2013.
Embassy of Socialist Republic of Vietnam in the USA. 2013. Constitution and political system. http://vietnamembassy-usa.org/vietnam/politics. Accessed 11 June 2013.
Hirsch, Philip, Kurt Mørck Jensen, Ben Boer, Naomi Carrard, Stephen FitzGerald, and Rosemary Lyster. 2006. National Interests and Transboundary Water Governance in

the Mekong. http://sydney.edu.au/mekong/documents/mekwatgov_mainreport.pdf. Accessed 23 July 2011.

International Rivers. 2011. The Xayaburi Dam: A Looming Threat to the Mekong River. www.internationalrivers.org/files/The%20Xayaburi%20Dam_Eng.pdf. Accessed 20 April 2011.

—— 2013a. Xayaburi Dam. www.internationalrivers.org/campaigns/xayaburi-dam. Accessed 28 March 2013.

—— 2013b. Xayaburi Dam: Timeline of Events. www.internationalrivers.org/files/attached-files/xayaburi_dam_timeline_of_events_feb._2013_0.pdf. Accessed 28 March 2013.

—— undated. About International Rivers. www.internationalrivers.org/resources/about-international-rivers-3679. Accessed 26 March 2013.

IUCN. 2013. About IUCN. www.iucn.org/about/. Accessed 7 November 2013.

Kerkvliet, B.J.T. 2001. An approach for analysing state-society relations in Vietnam. *Sojourn: Journal of Social Issues in Southeast Asia* 16 (2): 238–278.

Law on Commune/Sangkat Administrative Management. 2001. Cambodia.

Law on Organization of the Government. No. 32/2001/QH10 of 25 December 2001. 2001. Vietnam.

Law on Vietnam Fatherland Front. No:14/1999/QH10 of 12 June 1999. 1999. Vietnam.

Law on Water Resources Management of the Kingdom of Cambodia. 2007. Cambodia.

McCargo, Duncan. 2005. Cambodia: Getting away with authoritarianism? *Journal of Democracy* 16 (4).

Middleton, Carl. 2012. Transborder environmental justice in regional energy trade in mainland South-East Asia. *ASEAS-Austrian Journal of South-East Asian Studies* 5 (2): 292–315.

MRC. 2006. Minutes of the twenty-fourth meeting of the Joint Committee, Mekong River Commission (29–30 August 2006, Vientiane, Lao PDR).

—— 2011. Report: Informal Donor Meeting, Mekong River Commission. 23–24 June 2011. Phnom Penh, Cambodia. www.mrcmekong.org/assets/Publications/governance/Minutes-of-IDM2011-final.pdf. Accessed 31 October 2011.

—— 2013a. Audited Financial Statements as at and for the year ended 31 December 2012. Mekong River Commission www.mrcmekong.org/assets/Publications/governance/2012-Audited-Reports-FS-FINAL.pdf. Accessed 8 April 2013.

—— 2013b. Joint Development Partner Statement: 19th MRC Council Meeting, 17 January 2013. ttp://www.mrcmekong.org/news-and-events/speeches/joint-development-partner-statement-19th-mrc-council-meeting-17-january-2013/. Accessed 4 May 2013.

—— undated-a. Development Partners & Partner Organisations. Mekong River Commission www.mrcmekong.org/about-mrc/development-partners-and-partner-organisations/. Accessed 11 July 2015.

—— undated-b. Organisational Structure. Mekong River Commission www.mrcmekong.org/about-the-mrc/organisational-structure/. Accessed 24 October 2012.

Nam, Minh, Tan Tu, and An Dien. 2012. Vietnam did not invade, but revived Cambodia: Hun Sen. *Thanh Nien News*, 5 January 2012. www.thanhniennews.com/index/pages/20120105-vietnam-did-not-invade--but-revived-cambodia--hun-sen.aspx Accessed 5 November 2013.

NGO Forum on Cambodia. 2011. List of Rivers Coalition in Cambodia (RCC). (A document obtained from an interviewee).

—— 2013. Who we are. www.ngoforum.org.kh/index.php/en/about-ngof/introduction. Accessed 21 January 2014.

—— undated. Hydropower and Community Rights Project. www.ngoforum.org.kh/eng/en_project_artticle.php?artticle=8. Accessed 13 October 2012.

Norlund, Irene. 2007. Filling the Gap: The Emerging Civil Society in Vietnam. Ha Noi. www.un.org.vn/en/component/docman/doc_details/3-filling-the-gap-the-emerging-civil-society-in-viet-nam.html. Accessed 14 April 2012.

Norlund, Irene, Dang Ngoc Dinh, Bach Tan Sinh, Dang Ngoc Quang, Do Bich Diem, Nguyen Manh Cuong, Tang The Cuong, and Vu Chi Mai. 2006. The Emerging Civil Society: An Initial Assessment of Civil Society in Vietnam. Vietnam Institute of Development Studies (VIDS). UNDP Vietnam. SNV Vietnam. Hanoi. www.civicus.org/new/media/CSI_Vietnam_report%20.pdf. Accessed 7 December 2011.

Obendorf, Robert B. 2004. Law Harmonisation in Relation to the Decentralisation Process in Cambodia. Working Paper 31. Cambodia Development Resource Institute. www.cdri.org.kh/webdata/download/wp/wp31e.pdf. Accessed 23 October 2011.

R3. 2012. Personal interview, 30 June 2012.

R8. 2012. Personal interview, 14 August 2012.

R10. 2011. Personal interview, 16 November 2011.

Rivers Coalition in Cambodia. 2012. Rivers Coalition in Cambodia (RCC) Terms of Reference (a document obtained from an interviewee).

—— undated. Rivers Coalition in Cambodia (RCC) Terms of Reference (a document obtained from an interviewee).

Save the Mekong. 2009. About Save the Mekong Coalition. www.savethemekong.org/issue_detail.php?sid=13. Accessed 19 October 2012.

—— undated. Save the Mekong Coalition. www.savethemekong.org/index.php?langss=en. Accessed 29 November 2010.

Sinh, Bach Tan. 2014. Identifying civil society in Vietnam. In *Southeast Asia and the Civil Society Gaze: Scoping a contested concept in Cambodia and Vietnam*, edited by G. Waibel, J. Ehlert and H. N. Feuer. London: Routledge.

Taylor, Robin, and MRC Secretariat. 2011. Mekong River Commission Annual Report 2010. www.mrcmekong.org/assets/Publications/governance/Annual-Report-2010.pdf. Accessed 28 March 2013.

TEAM Consulting Engineering and Management Co. Ltd. 2008. Feasibility Study Xayaburi Hydroelectric Power Project Lao, PDR. Final Report (Main Report). CH. Karnchang Public Company Limited. www.mrcmekong.org/assets/Consultations/2010-Xayaburi/xayaboury-dam-feasibility-study.pdf. Accessed 25 October 2011.

TERRA. 2008. Watershed: People's Forum on Ecology. Burma, Cambodia, Lao PDR, Thailand, Vietnam. Towards Ecological Recovery and Regional Alliance (TERRA). www.terraper.org/mainpage/watershed.php. Accessed 2 November 2011.

—— undated. Project for Ecological Recovery (PER) and Environmental Struggles: Some Perspectives on Thai Environmental Movement. www.terraper.org/file_upload/PartOne_PERPerspectives.pdf. Accessed 20 October 2012.

Thayer, C.A. 2008. One party rule and the challenge of civil society in Vietnam. In *Remaking the Vietnamese State: Implications for Vietnam and the Region, Vietnam Workshop*. Hong Kong (21-22 August 2008).

Un, Kheang. 2011. Cambodia: Moving away from democracy? *International Political Science Review / Revue internationale de science politique* 32 (5): 546–562.

UNDP Vietnam. 2006. Deepening Democracy and Increasing Popular Participation in Vietnam. UNDP Vietnam Policy Dialogue Paper 2006/1. www.vn.undp.org/content/

dam/vietnam/docs/Publications/4856_Grassroot_democracy.pdf. Accessed 15 November 2010.

United Nations Human Rights Council. 2010. Report of the Special Rapporteur on the situation of human rights in Cambodia. http://cambodia.ohchr.org/WebDOCs/DocReports/3-SG-RA-Reports/A_HRC_CMB16092010E.pdf. Accessed 26 May 2013.

V2. 2012. Personal interview, 10 July 2012.

V5. 2012. Personal interview, 11 July 2012.

V10. 2012. Personal interview, 14 July 2012.

V11. 2012. Personal interview, 14 July 2012.

V12. 2012. Personal interview, 16 July 2012.

V23. 2012. V23 E-mail communication with the author.

Vandenbrink, Rachel. 2013. Hun Sen says he will stay in power until he's 74. *Radio Free Asia*, 6 May 2013. www.rfa.org/english/news/cambodia/election-05062013185646.html Accessed 19 May 2013.

Vickery, Michael. 1984. *Cambodia 1975-1982*. Sydney: South End Press.

Viet Nam Farmers' Union. undated. Organization and operation of Vietnam Farmers' Union(VNFU).http://vnfu.vn/index.php/introduction/252-organization-and-operation-of-vietnam-farmers%E2%80%99-union-vnfu.html. Accessed 24 November 2013.

Vietnam Rivers Network. 2009. Vietnam Rivers Network's Strategy 2008-2020. www.warecod.org.vn/en-US/News/introduction/program/2010/05/212.aspx. Accessed 2 November 2011.

—— 2012. About Us/ Introduction. http://vrn.org.vn/en/h/d/2012/04/244/Introduction/index.html. Accessed 7 November 2012.

—— 2013. Báo Cáo Tông Kêt Hoạt Dông (Activity Report) 2006-2012. http://vrn.org.vn/media/files/VRN%20annual%20report_Final.pdf. Accessed 12 February 2013.

—— undated-a. About us: Network structure. http://vrn.org.vn/en/h/d/2012/04/246/Network_structure/index.html. Accessed 12 February 2013.

—— undated-b. Partners & Members/ Members. http://vrn.org.vn/en/h/d/2012/04/251/VRN's_Key_Members/index.html. Accessed 7 November 2012.

VUSTA. 2012a. About us. www.vusta.vn/english3/cate/?1604/General-Introduction.htm. Accessed 22 January 2014.

VUSTA. 2012b. *VUSTA Charter*. Hanoi: Vietnam Union of Science and Technology Associations.

—— undated. *The Vietnam Union of Science and Technology Associations*, hard-copy brochure. Hanoi: VUSTA.

6 Rules and norms

Introduction

This chapter provides an overview of the rules and norms that influenced the advocacy strategies of the two case-study coalitions. It first discusses rules and norms at the Mekong regional level, followed by a discussion of them at national levels. It also discusses rules and norms within the case-study NGO coalitions. The chapter discusses two main categories of rules and norms: 1) formal rules; and 2) informal rules and norms. This chapter does not include a discussion of rules and norms that did not seem to influence the NGOs, but focuses on those that this research has identified as exerting an influence over the case-study coalitions. The discussion also touches on the political and historical contexts associated with the development of rules and norms, where applicable.

Formal rules at the Mekong regional level

1995 Mekong Agreement

The 1995 Mekong Agreement, introduced in Chapter 4, is an agreement among four riparian states on the Mekong River. The Agreement aims to promote cooperation for the sustainable development, utilization, management, and conservation of the Mekong River Basin (Article 1 *Mekong Agreement* 1995). One of the main substantive rules of this agreement is reasonable and equitable utilization, which is outlined in Article 5. The principle of equitable and reasonable utilization (ERU) is the primary substantive rule that governs transboundary watercourses (Wouters *et al.* 2005: 21). In order to fulfil member states' obligations, the 1995 Mekong Agreement established the Mekong River Commission (MRC) as an institutional mechanism. As a way to implement substantive obligations, procedural rules were established. One procedural rule is the Procedures for Notification, Prior Consultation and Agreement (PNPCA), which forms the main process that this book examines and analyses.

Procedures for Notification, Prior Consultation and Agreement

As one way to operationalize the ERU principle, Article 5 of the 1995 Mekong Agreement requires states to notify and consult with other riparian states on the use of a river (Article 5 *Mekong Agreement* 1995). As a detailed procedural rule for this notification and consultation, the PNPCA was adopted by the MRC in 2003 (PNPCA 2003). Subsequently, the Guidelines on Implementation of the PNPCA were adopted by the MRC Joint Committee in 2005 (*PNPCA Guideline* 2005) in order to provide detailed guidelines for PNPCA implementation.

The 1995 Mekong Agreement and PNPCA require states to notify the Joint Committee of both intra-basin uses and inter-basin diversions in tributaries of the Mekong River, including Tonle Sap (the largest freshwater lake connected to the Mekong River Basin) (Article 5A *Mekong Agreement* 1995). The use of the mainstream water is also subject to notification and consultation, depending on the use and the season (Article 5B *Mekong Agreement* 1995). Prior consultation is required in addition to prior notification in the case of: 1) inter-basin diversion from mainstream during the wet season; 2) intra-basin use on the mainstream during the dry season; and 3) inter-basin diversion of the surplus quantity of water during the dry season (Article 5.1 PNPCA 2003). The Xayaburi dam is an intra-basin diversion during both dry and wet seasons, thus both notification and prior consultation are required.

While the 1995 Mekong Agreement is considered as a legally binding international treaty, there is a debate about whether or not procedural rules and guidelines developed under the Agreement are also considered as a treaty, and are therefore legally binding instruments (Rieu-Clarke 2015). This argument arises from the fact that the PNPCA and PNPCA guidelines are not ratified by the member states and may not meet the definition of a treaty defined by the Vienna Convention on the Law of Treaties (Rieu-Clarke 2015; *Vienna Convention* 1969). It is important to note this status since it is associated with perceptions of different actors over the controversial process of the Xayaburi dam, discussed further in Chapter 7.

However, irrespective of whether or not the PNPCA and PNPCA guidelines are legally binding *per se*, they offer valuable guidance in interpreting the provisions related to notification and consultation under the 1995 Mekong Agreement (Rieu-Clarke 2015; Article 31(3)(b) *Vienna Convention* 1969). The PNPCA provides detailed guidance on the content of notification and consultation including required studies, roles, and responsibilities of each actor involved in the process and time frame of the PNPCA (PNPCA 2003). PNPCA guidelines provide further interpretations of the details provided by the PNPCA (*PNPCA Guideline* 2005). For example, referring to the content of the notification, the PNPCA requires notifying states to provide a feasibility study, implementation plan, schedule, and all available data (Article 4.2.1 PNPCA). Referring to this article, PNPCA guidelines further note that most data shall be 'relevant available data' (Section I.A.3 *PNPCA Guideline* 2005), which is 'the data necessary for the notified parties to be informed of and to understand the proposed project and use of water to determine impacts upon them' (*PNPCA Guideline* 2005: n8). This point is particularly relevant to the Xayaburi dam PNPCA case, as the lack of transboundary impact assessment from the dam was debated among the states as well as by NGO and civil society actors during the PNPCA process (Kingdom of Cambodia 2011; Socialist Republic of Viet Nam 2011; Save the Mekong 2012).

Another important point to note is that the 1995 Agreement defines prior consultation as

> Timely notification plus additional data and information to the Joint Committee as provided in the Rules for water Utilization and Inter-Basin Diversion under Article 26, that would allow the other member riparians to discuss and evaluate the impact of the Proposed use upon their uses of water and any other affects, which is the basis for arriving at an agreement. Prior consultation is neither a right to veto the use nor unilateral right to use water by any riparian without taking into account other riparians' rights.
>
> (Chapter II *Mekong Agreement* 1995)

The wording associated with this requirement has more ambiguity compared with the requirement of mainstream water use under the Joint Declaration of Principles for Utilization of the Waters of the Lower Mekong Basin, signed in 1975 (*Mekong Joint Declaration* 1975). As discussed in Chapter 4, the 1975 Declaration considers mainstream waters as 'a resource of common interest not subject to major unilateral appropriation by any riparian State without prior approval by the other Basin States through the Committee' (Article X *Mekong Joint Declaration* 1975). This ambiguous nature of the requirement under the 1995 Agreement created difficulties during the PNPCA process of the Xayaburi dam, discussed in Chapter 7.

Stakeholder participation policies and plans adopted by the MRC

Since the 1995 Mekong Agreement is an agreement among states, the notification and consultation requirements do not specifically include the public as subject to be consulted (*Mekong Agreement* 1995; PNPCA 2003). Therefore the requirement stipulates only consultation with other member states (Section 1 PNPCA 2003). There are, however, policies and guidelines created through the MRC which support public participation in the governance of the Mekong River. One is a policy document adopted in 1999 called *Public Participation in the Context of the Mekong River Commission* (MRC 1999). The MRC strategy on public participation was developed in 2003 and an action plan for public participation was finalized in 2004 (MRC-BDP2 2009: 4). Stakeholder participation is also integrated into various programmes of the MRC (MRC 2005). In particular, two programmes with basin development plans in two different phases have both developed stakeholder participation plans (MRC 2004; MRC-BDP2 2009).

These MRC policies and plans concerning stakeholder participation provide detailed suggestions of how to identify stakeholders and engage them in the discussions associated with the development occurring in the Mekong River Basin (MRC 1999, 2004; MRC-BDP2 2009). Stakeholders are defined broadly by the MRC policy (1999) as

> any person, group(s) or institution that has an interest in an activity, project or programme. This includes both intended beneficiaries and intermediaries, those positively affected, and those involved and/or those who are generally excluded from the decision-making process.
>
> (MRC 1999: 2)

Touching on the key principles of public participation, the MRC policy indicates that participation needs to start as early as possible in a process and be transparent (MRC 1999). While the BDP2 Stakeholder Participation and Communication Plan accommodates specific elements of the BDP process, the technical elements of stakeholder engagement included in this plan could also be adopted by other initiatives within the MRC scheme.

Formal rules in Cambodia

Formal rules for interest articulation and decision-making

Cambodia is a constitutional monarchy (Article 1 *Constitution of Cambodia* 1993). The Constitution is an important formal instrument which provides the basis for human rights, including freedom of speech and freedom of association. The Constitution adopted in 1993 provides the legal basis for governing the country democratically, giving power to the people, who exercise their powers through the National Assembly, the Senate, the Royal Government, and the Judiciary (Article 51 *Constitution of Cambodia* 1993). Cambodian citizens have the right to vote directly in the elections of the National Assembly and the Commune council (Articles 76, 78 *Constitution of Cambodia* 1993). This offers an important opportunity and right for interest articulation because these elections provide opportunities for ordinary citizens to express their political preferences.

Formal rules on decentralization

Cambodia has a decentralized government system and adopts democracy and pluralism as its governing principle (Article 51 *Constitution of Cambodia* 1993). Through major support from international donors, the Cambodian decentralization process started in the early 2000s when two key laws were passed: the Law on the Administration and Management of Communes; and the Commune Election Law (*Law on Commune Management* 2001; *Amended*

Law on Elections of Commune Councils 2006; Öjendal and Sedara 2006: 510). The first commune council election was held in 2002 (Rusten *et al.* 2004: 14). This political system of decentralization created a situation for Cambodian political parties in which winning commune/*sangkat* elections is paramount to sustaining their political power at the national level (Öjendal and Sedara 2006: 510). The members of sub-national level (provincial and district) councils are elected by commune and *sangkat* council members (Article 27 *Law on Sub-National Council Election* 2008). At the national level, while National Assembly members are elected by general election, the majority of the Senate are elected by the members of the National Assembly and commune/*sangkat* council members (Article 11 *Law on Senate Election* 2005; Article 100 *Constitution of Cambodia* 1993)

Formal rules associated with freedom of expression

The Constitution of Cambodia guarantees personal freedom to any individual (Article 31 *Constitution of Cambodia* 1993). The rights of freedom of expression, press, publication, and assembly are all guaranteed in the Constitution (Article 41 *Constitution of Cambodia* 1993). The Law on the Press also aims to guarantee the freedom of the press (Article 1 *Law on the Press* 1995) and bans pre-publication censorship (Article 3 *Law on the Press* 1995). However, at the same time, the Law on the Press also restricts journalists from publishing information that harms someone's honour or dignity; this is used at times to punish journalists who criticize public figures (Article 10 *Law on the Press* 1995; Un 2011).

Formal rules governing NGOs in Cambodia

Prior to 1991, NGO presence in Cambodia was limited to emergency relief and support to refugees who were the victims of internal conflict in Cambodia. These NGOs were primarily international (Bañez-Ockelford 2010: 4). National NGOs in Cambodia emerged after the Paris Peace Agreement was signed in 1991, lifting the aid embargo imposed by the United Nations since 1982. Since 1991, local (Cambodian) NGOs increased significantly, primarily as a result of the availability of donor funding and the need to implement the agenda set by the donors (Bañez-Ockelford 2010: 4).

Administratively, local NGOs are required to register with the Ministry of the Interior (MoI) whereas international NGOs are required to register with the Ministry of Foreign Affairs (CCC 2012b). The registration process is relatively simple, and currently NGOs are widespread within Cambodian society (CCC 2012b). As of 2010, 1,591 local NGOs were registered with the MoI, and 508 international NGOs were registered with the Ministry of Foreign Affairs (CCC 2012a: 20).

Many local NGOs are engaged in development activities including education, health, credit, income-generation, and other rural and urban poverty-reduction initiatives (Bañez-Ockelford 2010: 6), often supplementing the work of the government. When conducting activities closely related to a particular government organization's mandate, NGOs often employ government staff as a 'counterpart' working alongside the particular NGO (C35 2012; C30 2012). If government staff are inside an NGO, it is difficult for the NGO to be critical about the government (C30 2012). NGOs are encouraged to take part in commune and district council meetings, particularly during the district planning workshops which are part of the local planning process (Bañez-Ockelford 2010: 5). NGOs with the main objective of advocacy work are a minority within the NGO community in Cambodia; they consist of only 7 per cent of the NGOs in Cambodia (Bañez-Ockelford 2010: 16).

The legal requirements relating to the operation of NGOs are referred to in the Civil Code, providing a relatively straightforward process of registering NGOs (Article 49 *The Civil Code of Cambodia* 2008). As a result of this situation, there are many civil society organizations (CSOs) that are registered but are not in operation. According to a census by the Cooperation Committee for Cambodia (CCC), as of 2010 only 30 per cent of CSOs registered with the MoI were active (CCC 2012a). According to the CCC, CSOs in Cambodia, which include NGOs, have been operating relatively freely (CCC 2012b).

In order to tighten the control over CSOs, the Cambodian government announced its intention to introduce the Law on Associations and Non-Governmental Organizations (LANGO) in 2008, and since then has produced several versions of this draft law (CCC 2012b). Compared with the current Civil Code, the proposed new law requires a cumbersome process of registration, making it almost impossible for some of the organizations to register formally (CCC 2011; *The Civil Code of Cambodia* 2008). In 2011, three UN Special Rapporteurs (the Special Rapporteur on the situation of human rights defenders; the Special Rapporteur on the rights to freedom of peaceful assembly and of association; and the Special Rapporteur on the situation of human rights in Cambodia) sent a joint letter to the government pointing out the importance of easier access to the NGO registration process and of the government guaranteeing the right for associations to appeal against a refusal of registration (UNHRC 2011: para. 28).

Formal rules governing the Rivers Coalition in Cambodia

The RCC's Terms of Reference (ToR) determine the way in which decisions are made within the RCC (RCC undated). According to the ToR, activities of the RCC are discussed at bimonthly meetings which all members are invited to attend. The RCC operates on a consensus basis. In case consensus cannot be reached among all members, or there is an urgent issue that needs deciding, certain members designated as 'core members' take the responsibility for decision-making on behalf of the network (RCC undated: 19; C3 2012). During the fieldwork conducted for this research, two versions of ToRs were obtained: an older version which is undated, and a new version drafted in October 2012. Both ToRs were consulted during the study (RCC undated,

RCC 2012). However, as this book reviewed advocacy activities that took place from September 2010 until August 2012, it refers primarily to the older version of the ToR when analysing the influence of this formal rule.

Informal rules and norms in Cambodia

Neo-patrimonialism, fear towards authorities, and taboo in criticizing others

Historically Cambodia has been a monarchical state ruled by its king. Through this rule, Cambodian society traditionally has characteristics of patrimonialism (Pak *et al.* 2007: 49). Patrimonialism is 'a power regime based on the personal power of the patron, and his/her discretionary ability to dispense favour and resources to clients, who in return rule as sub-patrons within their own domains' (Pak *et al.* 2007: 63). Khmer society traditionally adopted patronage at family, kinship, and village levels (Pak *et al.* 2007: 53). This Khmer tradition is backed by a belief of Theravada Buddhism that people who have a high status and great power are generally believed to have done good deeds in their past lives (Rotha and Vannarith 2008: 7; Pak *et al.* 2007: 42). This contributes to people's acceptance of current power relationships between 'rulers' and 'ruled'. It is also taboo to criticize someone in public, even if it is a constructive criticism (Rotha and Vannarith 2008: 10). Conflict avoidance and 'saving face' is another important cultural aspect of Cambodian citizens' relationships with their authorities. The hierarchical nature of Cambodian society does not allow people who are at a 'lower level' in the society to criticize people who are more powerful and occupy higher-level positions in society (Pak *et al.* 2007: 55). These cultural contexts contribute to creating 'fear' towards authorities. Öjendal and Sedara (2006) refer to '*korob* (respect), *kaud* (admiration) and *klach* (fear)' as key concepts representing Khmer citizens' general feelings towards local authorities (Öjendal and Sedara 2006).

Traditional patrimonialism based on kinship ties within a community transformed into neo-patrimonialism, which helped formal bureaucratic institutions gain power during the French colonial regime of the 1860s (Pak *et al.* 2007: 50, 57). Together with neighbouring Vietnam, Cambodia was a French colony from 1863 to 1953 (Cambodian Tribunal Monitor undated). The French regime hired Cambodian civil servants who were given prestige by the colonial regime and were in a position to extract taxes from rural peasants, without having accountability to local people (Pak *et al.* 2007: 50; Rusten *et al.* 2004: 87). In 1953, Cambodia gained its independence from France. The political regime that followed independence in 1953 intensified the power of patronage through neo-patrimonialism where the 'chief executive maintains authority through personal patronage rather than through ideology or law' (Pak *et al.* 2007: 63; Bratton and Van de Walle 1994: 458). This helped establish an authoritarian and corrupt political environment (Pak *et al.* 2007: 50). The Khmer Rouge, who ruled Cambodia from 1975 to 1979, used power in extreme ways, controlling food supplies and using violence (Pak *et al.* 2007: 52;

Ledgerwood *et al.* 1994: 12). Under an ideology based on the desire to build an economy based on agricultural collectivism, the Khmer Rouge forced both urban and rural citizens into forced labour conditions, without providing sufficient food for them (Ledgerwood *et al.* 1994: 12). Authorities under the People's Republic of Kampuchea, which came into being after the notorious Khmer Rouge period, also confiscated surplus harvests from poor people and forced them into collective work. This continued to promote the existing hierarchy system and did not allow Cambodians to question the authorities (Öjendal and Sedara 2006: 513; Pak *et al.* 2007: 52).

The Khmer Rouge significantly affected kinship and social ties in Cambodian society through its policies of separating family members and creating distrust among individual family members. This still affects Cambodian society (Pak *et al.* 2007: 57). Children were separated from their parents and husbands and wives were separated (Ledgerwood *et al.* 1994: 13). People were encouraged to spy on each other, and some children were taught to spy on their parents (Clarke 1993: 66; Rotha and Vannarith 2008: 8). Although the Khmer Rouge regime ended in 1979, many interviewees indicated that spies commissioned by the ruling Cambodian People's Party (CPP) still exist at various levels within the society, including among university students, monks, moto-dops (motorbike taxi drivers), and journalists (C25 2012; C35 2012; C21 2012). Under the current system, the CPP pays a small amount of money to these 'spies' who, in return, report any activities that could be potential threats to the ruling party (C21 2012; C25 2012; C35 2012). This behaviour creates distrust and fear in Khmer society, another factor contributing to the informal pressure exerted by the authorities.

The Khmer Rouge era created significant scars in Cambodian society, destroying families and communities. In 1979, the Vietnamese army, along with some former Khmer Rouge members, invaded Cambodia, establishing the People's Republic of Kampuchea (PRK). While the new regime, established by Vietnam and led by Heng Samrin, ruled the country, the Khmer Rouge forces still fought to regain power. This resulted in another decade of civil war (Vickery 1984). The Paris Peace Agreement, signed in 1991, officially ended the internal conflict in Cambodia. The country entered a period of democratic regime which officially granted rights to citizens for political participation (Ledgerwood *et al.* 1994: 16). The introduction of this new system of governance could have allowed Cambodia to establish a non-authoritarian, democratic state with leaders selected through an elective process. However, neo-patrimonialism continued to develop during this period, when the CPP effectively used a 'carrot-and-stick' policy to strengthen their power and to win the hearts of rural voters (Un 2011: 556).

Starting in 1998, the CPP's Party Working Groups (PWG) created patronage networks from central to local levels, often operating in parallel to state institutions, and channelling centrally controlled resources into rural infrastructure development in the communities (Hughes 2009: 159; Un 2011: 556). At times, this system created the situation where communities supporting the opposition parties did not receive funds that should have been allocated to their communities' development (C30 2012; C35 2012), creating pressure within the village to vote for the CPP.

The CPP party members continue to receive favours. At village level, CPP members receive more support when there is a land dispute or competition over natural resource use (C30 2012; C35 2012). At times, community members who are labelled as non-CPP members could face discrimination within their community (Rotha and Vannarith 2008: 5). Government staff members are expected to provide support to the CPP's political campaigns (C30 2012). In addition, nepotism is rampant in the government system, allowing family members of high-ranking officials to be promoted, and giving CPP party members an additional advantage (C30 2012; Global Witness 2007). Corruption occurs widely at every level in Cambodia (Un 2006). Despite international and domestic criticism about corruption, according to Un (2006), no high-ranking officials have ever being punished for corruption. Thus there is zero risk in continuing corruption (Un 2006: 229). The personal advantage that results from supporting the CPP brings incentives for ordinary citizens to support the CPP and discourages them from disobeying the party.

There has also been a major crackdown on the opposition parties. The most symbolic event was in 1997, when Hun Sen orchestrated a *coup-d'état* that resulted in ousting Prince Norodom Ranariddh, the leader of the ruling FUNCINPEC party (Front uni national pour un Cambodge indépendant, neutre, pacifique, et coopératif: National United Front for an Independent, Neutral, Peaceful, and Cooperative Cambodia), and killing 40 top FUNCINPEC officers (Un 2011: 549). The government security personnel also attacked peaceful demonstrations organized by Sam Rainsy, the key opposition party leader, killing 16 people and wounding over 100 (Un 2011: 549). The election in 1998 resulted in the victory of the CPP. However, it was perceived by international observers and urban residents in Cambodia to be seriously flawed. Monks and students started peaceful demonstrations, demanding democracy and peace. The demonstration continued for approximately six months, and was joined by approximately 10,000 people (New York Times News Service 1998; C21 2012). However, it ended in bloodshed as the police and the military forcefully cracked down on demonstrators (C21 2012). This event silenced any further movements by monks and students in Cambodia. During the government's crackdown in 1998, a number of students disappeared, creating fear in other students and an unwillingness to speak up (C21 2012; C35 2012). Many monks have also disappeared and their bodies discovered later without clear explanations (C21 2012; C35 2012). All the Buddhist temples were 're-educated' by the ruling party to follow party ideology (C21 2012; C25 2012). Currently the head monk in Phnom Penh is appointed by the CPP and local Buddhist ceremonies are often used as part of the CPP's election campaigns (C25 2012; C9 2012).

The CPP's violent political crackdown resulted in an increase in opposition parties' popularity in the late 1990s (Un 2011: 556). However, the silencing of the opposition continues through the use of formal rules. The major opposition party leader, Sam Rainsy, has faced a number of criminal charges. In a recent case, in January 2010, Sam Rainsy was charged for racial incitement for simply removing some of the border posts installed on the border of Cambodia and Vietnam. He was sentenced to two years in prison according to the Criminal Law and Procedure of the United Nations Transitional Authority in Cambodia (UNTAC) (Human Rights Watch 2010; Anstis 2012: 317; *UNTAC Criminal Code* 1992). Later in the same year he was sentenced to ten years in prison for disinformation and falsification of public documents after he was charged with manipulating a map to demonstrate that Vietnam had encroached on Cambodian territory (LICADHO 2010; UNHRC 2011: 9). Surya Subedi, the United Nations special rapporteur on human rights in Cambodia, highlighted that this was an example of rule of law in Cambodia where courts implement laws in ways that do not conform to the scope of the law. Subedi criticized this sentence as politically motivated (UNHRC 2009: para 23). He continued:

> In any properly functioning democracy, such political matters would be debated in the parliament and become a matter of public debate rather than the subject of a criminal case before courts.
>
> (UNHRC 2009: para. 23)

The way that Cambodia uses law as a means of intimidating the opposition is now widely regarded as a practice adopted by authoritarian regimes to maintain control (Anstis 2012: 312).

Formal rules in Vietnam

Formal rules for interest articulation and decision-making

Vietnam is a socialist republic based on Marxism–Leninism and Ho Chi Minh's thoughts (*Constitution of Vietnam* 2001). Ho Chi Minh was President of the Democratic Republic of Vietnam (North Vietnam) from 1945 to 1969, and he is considered to be the founding father of modern Vietnam. The Constitution, adopted in 1992, provides the legal basis for Vietnam's political structure (*Constitution of Vietnam* 2001). The Constitution also guarantees the rights of people to be nominated and vote in elections to the National Assembly and the People's Councils, which play executive roles at national and local levels within Vietnam (Article 54 *Constitution of Vietnam* 2001). Citizens' rights to participate in the management of the state and send petitions to People's Councils are also rights guaranteed in the Constitution (Article 53 *Constitution of Vietnam* 2001). These are important rights for interest articulation for Vietnamese citizens.

Doi Moi

Vietnam is a single-party state, led by the Communist Party of Vietnam (CPV). The CPV was established in 1930 by Ho Chi Minh. The country became a battlefield during the Cold War from 1955–75 during the fight between the communist-supported North Vietnam and the US-supported South. The war ended in 1975 when the US army left South Vietnam. The entire country has been governed by the communist party since then.

During the Cold War, the CPV had close relationships with the Soviet Union and communist countries in Eastern Europe. The country also adopted a planned economy where corporations were state-owned and farmers were grouped into collective agricultural production units (Irvin 1995). This planned economy did not lead to a productive economy, resulting in the Vietnamese government's introduction of reformed economic policy, *Doi Moi*. *Doi Moi* decentralized state-owned corporations, allowed private enterprises to operate, and privatized agricultural production (Beresford 2008; Gainsborough 2010; Irvin 1995). This reform, which coincided with the collapse of Eastern European regimes, reoriented Vietnam's foreign relations from its earlier focus on communist bloc countries, to non-communist countries (Norlund *et al.* 2006: 10). For instance, in 1994 the USA lifted its trade embargo with Vietnam, which had been in place since the end of the Vietnamese war in 1975 (Cockburn 1994). This shift has also resulted in western donors increasing their development cooperation with Vietnam, resulting in an increase in the number of international NGOs present in the country, as well as an increase in local NGOs (Norlund *et al.* 2006: 10). The *Doi Moi* reform created an important shift in Vietnamese civil society.

Formal rules associated with freedom of expression

Both the Constitution of Vietnam and the Law on Media guarantee freedom of speech and freedom of the press (Article 69 *Constitution of Vietnam* 2001; Article 2 *Law on Media* 1999). These rights are, however, threatened at times through the use of a penal code which states:

> freedom of speech, freedom of press, freedom of belief, religion, assembly, association and other democratic freedoms which infringe upon the interests of the State, the legitimate rights and interests of organizations and/or citizens, shall be subject to warning, non-custodial reform for up to three years or a prison term of between six months and three years.
>
> (Article 258 *Penal Code, Vietnam* 1999)

The Law on Media guarantees citizens' rights to freedom of speech by ensuring their rights to contact and provide information to media without being subject to censorship, and to express opinions on domestic and world and current affairs (Article 4 *Law on Media* 1999). On the other hand, the same law obliges journalists to 'protect the guidelines and policies of the Party and the laws of the State; to seek out and protect positive initiatives; to fight against wrong ideology' (Article 15(2)(b) *Law on Media* 1999), which appears to counteract the freedom of expression guaranteed in the same law.

Formal rules governing NGOs in Vietnam

Many scholars discuss the close relationship between the state and civil society in Vietnam, indicating that the boundary between the state and civil society is fuzzy (Kerkvliet 2001: 241; Thayer 2008: 5-11; Norlund *et al.* 2006: 32). Historically, prior to 1986 when Vietnam adopted the *Doi Moi* policy, 'civil society' in the Vietnamese context primarily referred to 'mass organizations' (Norlund *et al.* 2006). As discussed in Chapter 5, mass organizations, *de facto*, work as the civil society arm of the CPV.

Organizations currently called 'Vietnamese NGOs' (VNGOs) emerged in the 1990s (Norlund *et al.* 2006: 26). As opposed to mass organizations that are associations of certain professions, the VNGOs are primarily issue-based organizations, and many focus their work on social and human development (Norlund *et al.* 2006). Initially, some of these VNGOs were established as science and technology organizations under Decree 35-HDBT (1992) on the establishment of non-profit and science and technology organizations, and Decree 81/2002/ND-CP on the implementation of the Science and Technology law (*Decree No. 35/HTBT* 1992; *Decree No. 81/2002/ND-CP* 2002; Norlund *et al.* 2006: 72).

Decree 88 was adopted in 2003 as a legal framework which allowed for the establishment and registration of 'associations', which distinguish VNGOs from mass organizations (*Decree No. 88/2003/ND-CP* 2003). While the word 'NGO' is currently used in Vietnam to refer to associations established under Decree 88, the word was not initially accepted within Vietnamese society. The Vietnamese word for NGO (*tô chúc phi chinh phu*) means 'organization external to the State', and in the Vietnamese context anything external to the state is viewed as anti-state or anarchist (Thayer 2008: 9; Norlund *et al.* 2006: 30).

The establishment of Vietnamese NGOs is permitted by different state institutions depending on the geographical scope of the NGO (Article 15 *Decree No. 88/2003/ND-CP* 2003). Decree 30/2003/ND-CP regulates operations of social and charity funds, including those of NGOs. One of the requirements under this decree is for NGOs to operate under government agencies with recognized competences (Article 4 *Decree No.30/2012/ND-CP* 2012). This article, by its nature, makes NGOs susceptible to government influence. With this requirement, many Vietnamese NGOs are registered under the Vietnam Union of Science and Technology Associations (VUSTA) (Norlund 2007: 11). The Vietnam Rivers Network (VRN), which is one of the case study NGO coalitions featured in this book, is registered as one of the projects of the Centre for Water Resources Conservation and Development (WARECOD), a Vietnamese NGO registered under VUSTA.

The Prime Minister's decision 22/2002/QD-TTg assigns VUSTA to objectively review the Vietnamese government's policies (PM decision 22/2002/QD-TTG). Through this mandate, VUSTA is able to provide its comments on hydropower dam projects to the Vietnamese government. This formal rule contributed to enhancing VUSTA's role in creating access to decision-makers over the issue of hydropower development on the Mekong, which the VRN is concerned about.

The regulation of VUSTA defines rights and obligations of VUSTA member associations (Article 7-9 *VUSTA Regulation* 2006). VNGOs registered under VUSTA are considered as their member associations (V7 2012). VUSTA's regulations define the rights of VUSTA member associations to participate in VUSTA activities (Article 9 *VUSTA Regulation* 2006).

The government promulgated a regulation on the management and use of foreign nongovernmental aid in 2009 (*Regulation on the use of aid from INGO* 2009). This regulation requires government agencies to approve the use of funds provided by foreign nongovernmental organizations. The government agencies that approve the use of funds depends on the amount and the purpose of funds (*Regulation on the use of aid from INGO* 2009).

Formal rules governing the Vietnam Rivers Network

The regulation of the VRN was adopted in 2009. It defines the structure of the VRN and the different roles assigned to each position within the network. One interesting aspect of the regulation is that it does not provide a clear mechanism for decision-making within the network. Instead, the regulation defines how and when the network's name can be used as members conduct activities. The regulation indicates that at least one member of the executive committee needs to participate in monitoring the use of the VRN's name (Article 21 VRN 2009). According to one of the interviewees, the decision-making process is not strictly regulated as the aim is to maintain flexibility in working style (V10 2012). The interviewee also claimed that since network members participate in various activities voluntarily, it is better not to make too strict rules (V10 2012).

Decisions related to the VRN's advocacy on the Xayaburi dam were determined within the VRN's Mekong task force, and the rest of the network would hear about the key issues or summary of the task force's activities through regular emails (V2 2012; V11 2012). This style of working within the VRN allowed the Mekong task force to conduct a variety of activities swiftly, compared with the way network decisions are made within the RCC (discussed in Chapter 9).

Informal rules and norms in Vietnam

This section considers the informal rules and norms that exist in Vietnamese society.

Valuing science

Vietnam has a long tradition of assigning importance to education and science. From 111 BC to 938 AD China ruled over most of current Vietnam, and this influenced Vietnam to adopt Confucian philosophy, providing an emphasis on education (Zink 2009: 9). The Vietnamese dynasties that ruled the country after the end of Chinese rule also emphasized education, symbolized by the Ly dynasty (1009–1225) which built the Temple of Literature in Hanoi (1070) (Zink 2009: 10). In reply to Vietnamese who demanded access to education, the French colonial government also made various arrangements for improving education, including founding the University of Indochina in Hanoi in 1906 (Zink 2009: 12), now the Vietnam National University in Hanoi.

Personal trust

Vietnam, as a traditional agrarian society with influences from Confucianism, traditionally has a tendency to trust small circles of acquaintances, particularly those consisting of family or village members (Turner and An Nguyen 2005: 1703; Dalton *et al.* 2001: 7). According to the World Values Survey conducted in 2001 by the Institute of Human Studies in Vietnam, 41 per cent of the respondents answered that people can be trusted, and the remaining 59 per cent answered that one needs to be careful when dealing with other people (Dalton *et al.* 2001: 6).

Fence-breaking

The Vietnamese concept of 'fence-breaking' means violation of rules and regulations that are set up by the party and the state (Vasavakul 2003). Traditionally, Vietnamese society had strong village communities which often played important roles in determining how centrally driven policy should be implemented at the local level (Lucius 2009: 8; Dalton and Ong 2005: 2). At times, Vietnamese citizens relied on 'fence-breaking' strategies to convey their policy preferences (Vasavakul 2003: 32).

Conclusion

This chapter provides the background to the rules and norms that affect the advocacy strategies of the two case-study NGO coalitions. This background is essential for the analysis of these strategies discussed in Part III. The comparison of rules and norms conducted in this chapter illustrates some key points that support the further analysis.

First, the chapter highlights differences in the formal rules that govern NGOs in Cambodia and Vietnam. In Vietnam, formal rules create stronger control of NGOs by the state, compared with Cambodia. This is an important factor when considering state–NGO relationships in both countries, discussed further in Chapter 9.

Second, the analysis in this chapter highlights the influence of external donors in shaping NGOs in both Cambodia and Vietnam. The emergence of the NGO sector in Cambodia was a result of the availability of external donors providing funds for reconstruction of the country. Vietnamese NGOs emerged as a result of Vietnam opening its economy to the West through the *Doi Moi* policy.

Third, the comparison reveals differences in the formal rules governing the NGO coalitions themselves. While the RCC has a clear and formalized decision-making mechanism, the VRN's decision-making mechanism is not formalized. The difference in these formal rules also affects the working culture within each coalition and influences strategies. This is discussed in more detail in Part III.

Finally, the comparison identifies that the constitutions of both Cambodia and Vietnam guarantee freedom of expression. However, other laws and formal rules that exist within both countries have Articles that undermine this right, particularly for the media. This contradictory relationship between formal rules has an important influence on NGO strategies towards the media. This is discussed in Chapter 11.

References

1992 Constitution of the Socialist Republic of Vietnam: As Amended 25 December 2001 (transl. Allens Arthur Robinson). 2001. Vietnam.

Agreement on the Cooperation for the Sustainable Development of the Mekong River Basin. 1995.

Amended Law on Elections of Commune Councils. 2006. Cambodia.

Anstis, Siena. 2012. Using law to impair the rights and freedoms of human rights defenders: a case study of Cambodia. *Journal of Human Rights Practice* 4 (3): 312–333.

Bañez-Ockelford, Jane. 2010. *Reflections, Challenges and Choices: 2010 Review of NGO Sector in Cambodia*. Phnom Penh: Cooperation Committee for Cambodia. www.ccc-cambodia. org/downloads/publications/2010_Review_NGO_Sector_Assessment.pdf. Accessed 7 June 2015.

Beresford, Melanie. 2008. *Doi Moi* in review: the challenges of building market socialism in Vietnam. *Journal of Contemporary Asia* 38 (2): 221–243.

Bratton, Michael, and Nicolas Van de Walle. 1994. Neopatrimonial regimes and political transitions in Africa. *World Politics* 46 (4): 453–489.

C3. 2012. Personal interview, 25 July 2012.

C9. 2012. Personal interview, 28 July 2012.

C21. 2012. Personal interview, 7 August 2012.

C25. 2012. Personal interview, 8 August 2012.

C30. 2012. Personal interview, 12 August 2012.

C35. 2012. Informal conversations with the author, September–October 2012.

Cambodian Tribunal Monitor. undated. *Cambodian History*. www.cambodiatribunal.org/ history/cambodian-history. Accessed 9 July 2013.

The Civil Code of Cambodia 2008. Cambodia.

Clarke, Greg. 1993. Three forms of stress in Cambodian adolescent refugees. *Journal of Abnormal Child Psychology* 21 (1): 65–77.

Cockburn, Patrick. 1994. US finally ends Vietnam embargo. *The Independent*, 4 February. www.independent.co.uk/news/world/us-finally-ends-vietnam-embargo-1391770. html. Accessed 10 July 2013.

The Constitution of the Kingdom of Cambodia. 1993. Cambodia.

CCC. 2011. *Consolidated Report on Issues and Recommendations on Draft Law on Associations and NGOs in Cambodia*. Phnom Penh: Cooperation Committee for Cambodia. www.ccc-cambodia.org/downloads/ngolaw/recommendations/Consolidated%20Report%20 on%20Issues%20and%20Recommendations%20on%20draft%20NGO%20Law%20 6-1-2011%20English.pdf. Accessed 16 June 2012.

—— 2012a. *CSO Contributions to the Development of Cambodia 2011*. Phnom Penh: Cooperation Committee for Cambodia. www.ccc-cambodia.org/downloads/ publications/CSO_Contributions_2011.pdf. Accessed 7 June 2015.

—— 2012b. Law on Association and Non-Governmental Organizations (LANGO). www. ccc-cambodia.org/lango.html. Accessed 16 June 2012; no longer available online.

Dalton, R. J., and N. N. T. Ong. 2005. Civil society and social capital in Vietnam. In *Modernisation and Social Transformation in Vietnam: Social Capital Formation and Institution Building*, edited by Gerd Mutz and Rainer Klump. Hamburg: Institut für Asienkunde.

Dalton, Russel J., Pham Minh Hac, Pham Thanh Nghi, and Nhu-Ngoc T. Ong. 2001. *Social Relations and Social Capital in Vietnam: The 2001 World Values Survey*. www. democracy.uci.edu/files/democracy/docs/vietnam/vietnam02.pdf. Accessed 2 June 2013.

Decree 30/2012/ND-CP of 12 April 2012 of the Government on the organization and operation of social funds and charity funds. 2012. Vietnam.

Decree 35/HTBT of 28 January 1992 of the Council of Ministers on Establishment of non-profit and science and technology organizations. 1992. Vietnam.

Decree 81/2002/ND-CP of 17 October 2002 of the Government on detailing the implementation of a number of articles of the science and technology law. 2002. Vietnam.

Decree 88/2003/ND-CP of 30 July 2003 of the Government on providing for the organization, operation and management of associations. 2003. Vietnam.

Gainsborough, Martin. 2010. *Vietnam: Rethinking the State*. London: Zed Books.

Global Witness. 2007. *Cambodia's Family Trees: Illegal logging and the stripping of public assests by Cambodia's elite*. www.globalwitness.org/sites/default/files/pdfs/cambodias_family_ trees_low_res.pdf. Accessed 28 October 2012.

Hughes, Caroline. 2009. *Dependent Communities: Aid and Politics in Cambodia and East Timor*. Ithaca, NY: Cornell Southeast Asia Program Publication.

Human Rights Watch. 2010. Cambodia: Opposition Leader Sam Rainsy's trial a farce. www.hrw.org/news/2010/01/28/cambodia-opposition-leader-sam-rainsy-s-trial-farce. Accessed 27 May 2013.

Irvin, George. 1995. Vietnam: assessing the achievements of *Doi Moi*. *Journal of Development Studies* 31 (5): 725.

Joint Declaration of Principles for Utilization of the Waters of the Lower Mekong Basin, signed by the representatives of the governments of Cambodia, Laos, Thailand and Vietnam to the Committee for Coordination of Investigations of the Lower Mekong Basin, signed at Vientiane on 31 January 1975.

Kerkvliet, Benedict J. Tria. 2001. An approach for analysing state–society relations in Vietnam. *Sojourn: Journal of Social Issues in Southeast Asia* 16 (2): 238–278.

Kingdom of Cambodia. 2011. *Mekong River Commission Procedures for Notification, Prior Consultation and Agreement: Form/Format for Reply to Prior Consultation*. Cambodia National Mekong Committee (CNMC). www.mrcmekong.org/assets/ Consultations/2010-Xayaburi/Cambodia-Reply-Form.pdf. Accessed 25 November 2011.

Law on Commune/Sangkat Administrative Management. 2001. Cambodia.

Law on Elections of Capital Council, Provincial Council, Municipal Council, District Council and Khan Council. 2008. Cambodia.

Law on Media. No. 12/1999/QH10 of 12 June 1999. Vietnam.

Law on Senate Election. 2005. Cambodia.

Law on the Press. 1995. Cambodia.

Ledgerwood, Judy, May M. Ebihara, and Carol A. Mortland. 1994. Introduction. In *Cambodian Culture Since 1975: Homeland and Exile*, edited by M. M. Ebihara, C. A. Mortland and J. Ledgerwood. Ithaca, NY: Cornell University Press.

LICADHO. 2010. *Cambodia Monthly News Summary* – September 2010. www.licadho-cambodia.org/printnews.php?id=126. Accessed 27 May 2013.

Lucius, Casey. 2009. *Vietnam's Political Process: How Education Shapes Political Decision-Making.* Hoboken, NJ: Routledge.

MRC. 1999. *Public Participation in the Context of the Mekong River Commission.* Mekong River Commission. www.mrcmekong.org/assets/Publications/policies/Public-Partici pation-in-MRC-context.pdf. Accessed 17 October 2011.

—— 2004. *The MRC Basin Development Plan: Stakeholder Participation.* BDP Library Volume 5. Mekong River Commission.

—— 2005. *Public Participation in the Lower Mekong Basin.* Mekong River Commission. www.mrcmekong.org/assets/Publications/governance/Public-Participation.pdf. Accessed 5 May 2012.

MRC-BDP2. 2009. *Stakeholder Participation and Communication Plan for Basin Development Planning in the Lower Mekong Basin.* Mekong River Commission: Basin Development Plan Programme Phase 2. www.mrcmekong.org/assets/Publications/Consultations/ BDP-POE-2010/SPCP-Final-July-2009-Final.pdf. Accessed 25 October 2011.

New York Times News Service. 1998. Thousands join protest in Cambodia. *Chicago Tribune* 11 September. http://articles.chicagotribune.com/1998-09-11/news/9809110 325_1_police-officers-cambodia-election-result. Accessed 13 October 2012.

Norlund, Irene. 2007. *Filling the Gap: The Emerging Civil Society in Vietnam.* Ha Noi. www. un.org.vn/en/component/docman/doc_details/3-filling-the-gap-the-emerging-civil-society-in-viet-nam.html. Accessed 14 April 2012.

Norlund, Irene, Dang Ngoc Dinh, Bach Tan Sinh, Dang Ngoc Quang, Do Bich Diem, Nguyen Manh Cuong, Tang The Cuong, and Vu Chi Mai. 2006. *The Emerging Civil Society: An Initial Assessment of Civil Society in Vietnam.* Hanoi: Vietnam Institute of Development Studies (VIDS)/UNDP Vietnam/SNV Vietnam. www.civicus.org/new/ media/CSI_Vietnam_report%20.pdf. Accessed 7 December 2011.

Öjendal, Joakim, and Kim Sedara. 2006. Korob, Kaud, Klach: in search of agency in rural Cambodia. *Journal of Southeast Asian Studies* 37 (3): 507.

Pak, K., V. Horng, N. Eng, S. Ann, S. Kim, J. Knowles, and D. Craig. 2007. *Critical Literature Review on Accountability and Neo-Patrimonialism: Theoretical Discussions and the Case of Cambodia.* Working Paper 34. Phnom Penh: Cambodia Development Resource Institute.

Penal Code. No. 15/1999/QH10. 21 December 1999. 1999. Vietnam.

PNPCA. 2003. *Procedures for Notification, Prior Consultation and Agreement.* Approved by the 10th meeting of the MRC Council on 30 November 2003.

PNPCA Guideline. 2005. *Guidelines on Implementation of the Procedures for Notification, Prior Consultation and Agreement.* Adopted 31 August 2005 by the 22nd meeting of the MRC Joint Committee in Vientiane, Lao PDR.

Provisions Relating to the Judiciary and Procedure Applicable in Cambodia during the Transitional Period. Decision adopted by the Supreme National Council of Cambodia on 10 September 1992. (UNTAC Criminal Code). 1992. Cambodia.

Regulation on management and use of foreign nongovernmental aid promulgated by Decree No. 93/2009/ND-CP of 22 October 2009 of the Government 2009. Vietnam.

Regulations of Vietnam Union of Science and Technology Associations. Issued jointly with the decision No. 650/QD-TTg, 24 April 2006 of the Prime Minister of the Socialist Republic of Vietnam 'Approval of the Regulations of Vietnam Union of S-T Associations' 2006. Vietnam.

Rieu-Clarke, Alistair. 2015. Notification and consultation procedures under the Mekong Agreement: insights from the Xayaburi controversy. *Asian Journal of International Law* 5 (1): 143–175.

RCC. 2012. Rivers Coalition in Cambodia: Terms of Reference. (Document obtained from interviewee.)

—— undated. Rivers Coalition in Cambodia: Terms of Reference. (Document obtained from interviewee.)

Rotha, Chan, and Chheang Vannarith. 2008. Cultural challenges to the decentralization process in Cambodia. *Ritsumeikan Journal of Asia Pacific Studies* 24: 1–16.

Rusten, Caroline, Kim Sedara, Eng Netra, and Pak Kimchoeun. 2004. *The Challenges of Decentralisation Design in Cambodia*. Phnom Penh: Cambodia Development Resource Institute. www.cdri.org.kh/webdata/pordec/decendesign.pdf. Accessed 21 May 2013.

Save the Mekong. 2012. Subject: Request for clarifications on the prior consultation for the Xayaburi dam. A letter addressed to Mr Hans Guttman, CEO, Mekong River Commission, 20 April.

Socialist Republic of Viet Nam. 2011. *Mekong River Commission Procedures for Notification, Prior Consultation and Agreement Form for Reply to Prior Consultation*. Viet Nam National Mekong Committee. www.mrcmekong.org/assets/Consultations/2010-Xayaburi/Viet-Nam-Reply-Form.pdf. Accessed 25 October 2011.

Thayer, C. A. 2008. One party rule and the challenge of civil society in Vietnam. In *Remaking the Vietnamese State: Implications for Vietnam and the Region, Vietnam Workshop*, Hong Kong, 21–22 August 2008. www.viet-studies.info/kinhte/CivilSociety_Thayer.pdf. Accessed 7 June 2015.

Turner, Sarah, and Phuong An Nguyen. 2005. Young entrepreneurs, social capital and *Doi Moi* in Hanoi, Vietnam. *Urban Studies* 42 (10): 1693–1710.

Un, Kheang. 2006. State, society and democratic consolidation: the case of Cambodia. *Pacific Affairs* 79 (2): 225–245.

—— 2011. Cambodia: Moving away from democracy? *International Political Science Review/Revue internationale de science politique* 32 (5): 546–562.

UNHRC. 2009. *Report of the Special Representative of the Secretary-General for Human Rights in Cambodia, Surya Subedi*. United Nations Human Rights Council. http://cambodia.ohchr.org/WebDOCs/DocReports/1-SR-SRSG-Reports/A-SR-CMB31082009E.pdf. Accessed 26 May 2013.

—— 2011. *Report of the Special Rapporteur on the Situation of Human Rights in Cambodia*. http://cambodia.ohchr.org/WebDOCs/DocReports/3-SG-RA-Reports/A-HRC-18-46_en.pdf. Accessed 26 May 2013.

V2. 2012. Personal interview, 10 July 2012.

V7. 2012. Personal interview, 11 July 2012.

V10. 2012. Personal interview, 14 July 2012.

V11. 2012. Personal interview, 14 July 2012.

Vasavakul, Thaveeporn. 2003. From fence-breaking to networking: interests, popular organizations and policy influences in post-socialist Vietnam. In *Getting Organized in Vietnam: Moving in and around the Socialist State*, edited by B. J. T. Kerkvliet, R. H. K. Heng, and D. W. H. Koh. Singapore: Institute of Southeast Asian Studies.

Vickery, Michael. 1984. *Cambodia 1975–1982*. Sydney: South End Press.

Vienna Convention on the Law of Treaties. Adopted on 23 May 1969, entered into force on 27 January 1980. 1155 United Nations Treaty Series 331 (Vienna Convention).

VRN. 2009. Regulations of Vietnam Rivers Network. http://vrn.org.vn/en/h/d/2012/04/254/How_can_I_register_for_VRN_membership/index.html. Accessed 18 February 2013.

Wouters, Patricia K., Sergei Vinogradov, Andrew Allan, Patricia Jones, and Alistair Rieu-Clarke. 2005. *Sharing Transboundary Waters: An Integrated Assessment of Equitable Entitlement: The Legal Assessment Model*. IHP-VI Technical Documents in Hydrology No. 74. Paris: UNESCO/International Hydrological Programme. http://unesdoc.unesco.org/images/0013/001397/139794e.pdf. Accessed 5 May 2014.

Zink, Eren. 2009. *Science in Vietnam: An Assessment of IFS Grants, Young Scientists, and the Research Environment*. MESIA Impact Studies Report No. 9. Stockholm: International Foundation for Science. www.ifs.se/IFS/Documents/Publications/MESIA%20reports/MESIA_9_Vietnam.pdf. Accessed 18 July 2013.

Part III

Strategies and interactions

7 The Xayaburi dam story

Introduction

Based on understandings of the Xayaburi dam's physical and material conditions, key actors, and legal requirements under the Procedures for Notification, Prior Consultation and Agreement (PNPCA) presented in earlier chapters, this chapter provides a detailed description of the PNPCA process of the Xayaburi hydropower dam. The main focus is on an understanding of key actors' positions, as they can potentially influence the advocacy strategies adopted by the case-study NGO coalitions. As discussed in Chapter 3, the focus of the case studies is on activities that took place during the two years from the time Laos officially notified the MRC of its intension to build the Xayaburi dam, thus instigating the PNPCA process (from September 2010 to August 2012). This chapter focuses on activities during those two years.

In discussing the activities of the various actors, the overall time frame is divided into three parts in order to highlight the shift in the actors' positions and the dynamics in their relationships throughout the PNPCA process. These three parts are: 1) from the beginning of the PNPCA process (September 2010) until the end of the initial six months, when the Joint Committee met to reach an agreement in April 2011; 2) after the April 2011 Joint Committee meeting to the Mekong River Commission (MRC) Council meeting in December 2011; and 3) January–August 2012, during which time Laos officially acknowledged the construction of the Xayaburi hydropower dam.

Tables 7.1 to 7.3 illustrate this timeline of events and the different actions taken by different actors in each period. This highlights the chronology and the reactions to key events by different actors. The strategies of the Rivers Coalition in Cambodia (RCC) and the Vietnam Rivers Network (VRN) are not discussed in detail in this chapter; they are the main focus of chapters 8–11. However, they are included in the tables to support the analysis of their strategies in the following chapters. Some ongoing activities are not specifically listed in the tables, such as informal meetings with various officials. Activities by the RCC and VRN are indicated in Tables 7.1 to 7.3 according to their different target audiences: activities targeted at regional and national decision-makers; stakeholders in potentially affected areas; and the general public. These typologies of target audiences provide a roadmap for the analysis of strategies which follows in chapters 8–11.

The initial six months of the PNPCA process

The PNPCA process for the Xayaburi hydropower dam was initiated on 20 September 2010 when the Lao National Mekong Committee informed the MRC of its intention to build the Xayaburi hydropower dam and submitted relevant documentation for the project (MRC 2011d; MRC Secretariat 2011: i). Subsequently, all the submitted documents were circulated to the MRC Joint Committee members in October 2010 (MRC Secretariat 2011: Annex 1A). These documents included a feasibility study, the environmental impact assessment (EIA) report, and the social impact assessment (SIA) report. While these documents were shared among MRC member states, they were not disclosed to the public until a much later stage in the PNPCA process (MRC 2011c: 2; R2 2011). The MRC Joint Committee (JC) established the PNPCA working group in October 2010 and conducted a series of meetings to determine and implement the PNPCA (MRC Secretariat 2011). Following the PNPCA rule which suggested that the time frame for prior consultation should be a minimum of six months, the initial goal of the JC was to conclude the PNPCA process by 22 April 2011 (Article 5.5.1 *PNPCA* 2003; MRC Secretariat 2011: i). During this six-month period the MRC commissioned a team of independent experts to conduct technical reviews of the Xayaburi hydropower dam. The expert panel recommended that further information on transboundary impacts and mitigation measures were both required before proceeding with the Xayaburi dam (MRC Secretariat 2011). The result of the review was published in March 2011 as the Prior Consultation Project review report (MRC Secretariat 2011).

In October 2010, the same month as the PNPCA process was initiated, the final report of the MRC-commissioned strategic environmental assessment (SEA) of 12 mainstream Mekong hydropower dam plans was published (ICEM 2010a). The report recommended ten years' deferral for all planned mainstream hydropower dams (ICEM 2010b: 140). It advised riparian countries to conduct further studies and prepare a 'Mekong mainstream plan' requiring a framework of zoning and safeguards against any future development, including the identification of areas that the countries wished to keep for future generations (ICEM 2010b: 137–139). When the feasibility study and the environmental and social impact assessments of the Xayaburi dam were conducted, the SEA's recommendation was not ready and was therefore not taken into consideration. However, the report did provide the basis for the main arguments used by the NGOs in the region, as well as the governments of Cambodia and Vietnam, in suggesting further studies before decisions were made on the Xayaburi dam (Socialist Republic of Viet Nam 2011; Save the Mekong 2011).

In responding to the commencement of the PNPCA process of the Xayaburi dam, the Save the Mekong (STM) Coalition sent a letter requesting the MRC to halt the Xayaburi PNPCA process, pointing out the inadequacies in the process, particularly in the engagement of the public and disclosure of Xayaburi project-related documents (Save the Mekong 2010).

As a way to garner public opinion related to the dam, official public consultations were conducted by respective National Mekong Committees (NMCs) during this initial six-month period, except in Laos, where a series of consultations were conducted during the SIA studies (MRC 2011c: 15) which took place prior to the commencement of the PNPCA. The CNMC and the VNMC each conducted two stakeholder consultation workshops in their respective countries. One was targeted at national-level stakeholders, including government agencies and NGOs based in capital cities; another was targeted at communities facing potential impact from the Xayaburi dam (MRC 2011c). The Thai National Mekong Committee (TNMC) conducted four stakeholder workshops: three held in localities along the Mekong River's mainstream which faced potential impact from the Xayaburi dam, one held in Bangkok, the capital city (MRC 2011c: 12). According to one of the interviewees engaged in organizing these consultation workshops, the MRC provided financial support to conduct two workshops in each country (C29 2011). However, Thailand decided to spend additional government funds to conduct additional workshops (C29 2011, 2012). The Lao National Mekong Committee (LNMC) did not hold any workshops during the PNPCA period. The LNMC maintained that the series of community-level workshops conducted as part of the SIA of the Xayaburi dam between 2007 and 2010 were sufficient in terms of engaging with citizens (MRC 2011c: 2).

These official consultations could have been used as opportunities for integrating the views of various stakeholders. Instead, they faced criticism from the NGOs in the region, primarily about the insufficient time allowed for the invitation of participants and for discussion, and limitations in the selection of the participants who were invited and in the information provided to them (Save the Mekong 2011; R2 2011). The Thai People's Network for Mekong, which consists of approximately 50 NGOs and local groups in Thailand, sent a letter to the CEO of the MRC in January 2011 requesting suspension of the PNPCA process as the public consultations were conducted in haste, and without clear understanding of the possible impacts (Thai People's Network for Mekong 2011). The STM Coalition sent a similar letter in January 2011 to the members of the MRC Council requesting a halt in the PNPCA process, in order to endorse the SEA report, and to conduct a credible process engaging the public (Save the Mekong 2011).

At the time of the public consultation, some of the project documents including the EIA and the SIA reports were not publicly distributed, and the feasibility study report was made available to the public only in mid-February 2011, one month after the public consultation had started, providing another reason for criticism (Save the Mekong 2011; R2 2011; R6 2012; MRC 2011c: 2). The consultation process was also criticized for not allowing citizens of Laos to have an opportunity to be consulted during the PNPCA process (HELVETAS

Table 7.1 Key activities during the initial six months of the PNPCA

Time	Key events	Key reactions by NGOs in the Mekong region	RCC activity	VRN activity
Sep 2010	Laos notified its intention to build the Xayaburi dam, initiating the PNPCA process			
Oct 2010	PNPCA process officially commenced (1 Oct)	Save the Mekong (STM) Coalition sent letter to MRC CEO asking to halt Xayaburi PNPCA (13 Oct) (R)	RCC signed letter drafted by STM Coalition (R)	VRN signed letter drafted by STM Coalition (R)
Oct 2010	MRC Secretariat released final SEA report of Mekong's mainstream hydropower dams	SEA report and recommendations from the study used by NGOs in their arguments as reference points (R)		
Nov 2010				VRN and Pan Nature co-hosted public dialogue on Mekong mainstream dams targeting Vietnam National Assembly members and VRN members (N)
Jan 2011	17th MRC Council meeting held	STM Coalition sent letter to MRC Council members requesting a halt of Xayaburi PNPCA process (R)	Some RCC members individually signed letter drafted by STM Coalition (R)	VRN signed letter drafted by STM Coalition (R)
Feb–Mar 2011	A series of PNPCA public consultations in Cambodia, Vietnam, and Thailand, organized by respective National Mekong Committees. Feasibility study released to the public. EIA and SIA reports not made public at this point (14 Feb)	WWF commissioned an expert review of the feasibility study and EIA from a fish and fisheries perspective (Mar) (R)	RCC members participated in both PNPCA public consultation meetings in Cambodia and distributed Xayaburi fact sheet (translation of International Rivers fact sheet) to participants. RCC members held side meeting with Cambodia National Mekong Committee (CNMC) during one consultation (N,S)	VRN members participated in PNPCA national consultation meeting. None of the members were invited to consultation held in Mekong Delta. VRN distributed Xayaburi fact sheet (translation of International Rivers fact sheet) to participants at both consultations (N, S)

Time	Key events	RCC activity	Key reactions by NGOs in the Mekong region	VRN activity
Mar 2011	Lao government and Xayaburi power company signed a concession agreement for the project	Some RCC members signed letter to Prime Ministers of Laos and Thailand (R) RCC wrote thank-you letter to CNMC for collaboration during the public consultation (N) Mekong for Life event held by an RCC member in Kratie province, involving floating down the Mekong River in inner-tubes (S, P)	263 NGOs from 51 countries signed letter to Prime Ministers of Laos and Thailand requesting them to cancel Xayaburi dam (R)	VRN signed letter to Prime Ministers of Laos and Thailand (R) VRN organized open dialogue on costs and benefits of Xayaburi dam and other dams on the Mekong's mainstream, targeting government officials, CSOs and scientists (N)
Apr 2011	MRC JC agreed to defer Xayaburi decision to Ministerial level		Bangkok Post revealed ongoing road construction of Xayaburi dam (17 April) (R)	VRN wrote letter of petition to key government offices, including Prime Minister, Government Office, MONRE, Ministry of Agriculture and Rural Development, VNMC, and VUSTA (N)

(Sources: International Rivers 2013; MRC 2011b, 2011c, 2011d, 2012b; Thai People's Network for Mekong 2011; Save the Mekong 2012; Baran *et al.* 2011; Trandem 2011; International Rivers 2011; 263 NGOs 2011).

Activities targeted at: (R) regional decision-makers; (N) national decision-makers; (S) stakeholders in potentially affected areas; (P) the general public.

Laos 2011). Although community consultations were held as part of the SIA, some of the critical information which would have been useful for the communities in developing their opinions had not been available at the time; this included the results of the SEA study, which was completed only in October 2010.

In March 2011, 263 NGOs from 51 countries signed a letter to the Prime Ministers of Thailand and Laos requesting cancellation of the Xayaburi hydropower dam (263 NGOs 2011). This international-scale advocacy activity was coordinated by international NGOs including International Rivers (IR), Both Ends, and the Focus on the Global South (Bank Information Center 2011). In the same month, the World Wide Fund for Nature (WWF) published a report reviewing the feasibility study and the EIA report from a fish and fisheries perspective; this was published in March 2011 (Baran *et al.* 2011).

CH. Karnchang's annual report 2010 indicated that preliminary construction work on the Xayaburi hydropower dam started in the second half of 2010 (CH. Karnchang 2011: 78), at the same time as initial discussions over the dam were taking place among the riparian nations. On 17 April 2011, two days before the MRC JC special session was scheduled to meet, the *Bangkok Post* reported that road construction leading to the Xayaburi hydropower dam by CH. Karnchang had already begun prior to the decision on the dam (*Bangkok Post* 2011c). Viraphonh Viravong, director general of Laos' Department of Electricity, Ministry of Energy and Mines, defended this construction by indicating to journalists that this road work was requested by both Xayaburi and Luang Prabang provincial authorities, and was for the development of the area regardless of the status of the dam (*Bangkok Post* 2011a).

At the end of the initial six months of the Xayaburi PNPCA, the MRC JC special session was held on 19 April 2011. At this meeting the member countries attempted to take a decision on the Xayaburi hydropower dam. There were different opinions and the JC members were unable to come to an agreement. While Laos insisted that it would take the comments of other countries into consideration in the project, and that the PNPCA process should be over, the other three member countries suggested needs for further impact assessment and consultations. The MRC member countries therefore agreed to defer the decision on the Xayaburi hydropower dam to the ministerial level (MRC 2011b).

The Cambodian government position during the initial six months of the PNPCA was included in the official reply to the MRC (Kingdom of Cambodia 2011). In this official reply, the Cambodian government expressed its concern about the lack of information, particularly on the transboundary impacts downstream of the dam. These included fisheries, flow change, sediment balance, erosion, ecosystem, agriculture, and livelihoods (Kingdom of Cambodia 2011). The government also expressed its concern over the PNPCA process. These concerns included the lack of time period for sufficient consultation and the lack of timely disclosure of project information to the public (Kingdom of Cambodia 2011). This reflects the concerns expressed by the NGOs throughout the initial six months of the PNPCA consultation period (NGO Forum on Cambodia 2011; Save the Mekong 2011). This official response was significantly different from the comment made by the Secretary General of the CNMC, Pich Dun, at the beginning of the PNPCA process, that the government had organized consultations with civil society in the past, and he was not sure about the need for further public consultation (Roeun and McGillian 2010).

As discussed earlier, the Vietnamese government's main concern was the Xayaburi dam's impact on the Mekong Delta, a region that supports the livelihoods of nearly 20 million people in Vietnam (Socialist Republic of Viet Nam 2011). In its official PNPCA reply to Laos, the Vietnamese government indicated that evidence of changes in the Mekong Delta had already been observed from the development in the upper reaches of the Mekong River, and suggested that all the hydropower projects on the Mekong mainstream be deferred for at least ten years, in order to allow sufficient time to conduct studies on cumulative impacts (Socialist Republic of Viet Nam 2011).

The Xayaburi scaling-up to a high political agenda

During the eight months following the MRC Joint Committee meeting in April 2011, the Xayaburi dam became an issue of concern at prime ministerial level in both Cambodia and Vietnam (Vrieze and Bopha 2011). The Xayaburi dam was the subject of discussion at high-level official meetings, many of which took place outside the MRC governance framework. As an example, during the meeting between the Prime Ministers of Vietnam and Cambodia shortly after the MRC JC meeting in April, the Prime Ministers confirmed the importance of taking the negative impacts of the dam into consideration (*Viet Nam News* 2011). During the side meeting of the eighteenth ASEAN summit on 7 May 2011, the Prime Minister of Laos, Thoongsing Thammavong, informed the Vietnamese Prime Minister Nguyen Tan Dung of the government's decision to temporarily suspend the Xayaburi hydropower dam project, a decision appreciated by Dung (Vietnam Plus 2011). However, on 8 June 2011 the Ministry of Energy and Mines of Laos issued a letter to the Xayaburi Power Company with its opinion that the PNPCA process had ended at the MRC JC level, since Laos had already provided the member countries of the MRC with opportunities for evaluating the Xayaburi project (Phomsoupha 2011). This highlighted the issue that whether the PNPCA process was complete or not was open to interpretation in different ways by different member states.

NGOs in the region reacted to the actions of Laos. The letter from the Lao Government to the Xayaburi Power Company was leaked to IR, which in turn commissioned Perkins Coie, a US law firm, to conduct a legal review to determine whether the Lao's conclusion regarding the status of the PNPCA was erroneous (Higgs 2011). Referring to the 1995 Mekong Agreement (Chapter II *Mekong Agreement* 1995), the review concluded that Laos could not

Table 7.2 Key activities May–December 2011

Time	Key events	Key reactions by NGOs in the Mekong region	RCC activity	VRN activity
May 2011	Lao Prime Minister indicated suspension of Xayaburi project while at eighteenth ASEAN summit (7 May)	STM Coalition released statement to eighteenth ASEAN summit calling for ASEAN leaders to urge cancellation of Xayaburi dam (R)		
Jun 2011	Laos' Ministry of Mines and Energy wrote to Xayaburi Power Company that the PNPCA process was completed	IR commissioned US law firm Perkins Coie to review Laos' unilateral termination of PNPCA (R)		
	During MRC Donor meeting in June Laos insisted that the PNPCA process was completed	MLN sent legal briefing note to MRC members (R)		
Jul 2011				VRN conducted workshop on hydropower dams targeting provincial governments and scientists in the Mekong Delta (S)
Aug 2011	Pöyry report evaluating Laos' compliance with 1995 Mekong Agreement published	IR and WWF separately published review of Pöyry report, heavily criticising the report's credibility (R)		
Oct 2011	Electricity Generating Authority of Thailand (EGAT) signed power purchase agreement with Xayaburi Power Company			

Time	Key events	Key reactions by NGOs in the Mekong region	RCC activity	VRN activity
Nov 2011			Xayaburi youth forum, conducted by RCC, invited Phnom Penh–based university students to discuss Xayaburi dam issue; statement of forum communicated to media (P) Xayaburi radio talk show organized by Culture and Environment Preservation Association (CEPA) (P)	One VRN Mekong task force member joined Vietnamese government's delegation for bilateral discussion with Laos VRN conducted workshop with VUSTA targeting scientists and media (N, P) VRN organized training workshop on sediments and geomorphology with WWF (N) VRN and ForWet (Research Centre for Forests and Wetlands) organized workshop for provincial governments in Mekong Delta with Southwest Steering Committee (S) VRN organized workshop for farmers in Mekong Delta to discuss impacts of hydropower dams (S)
Dec 2011	Eighteenth MRC Council meeting agreed further impact studies before Xayaburi dam could proceed	Decision welcomed by NGOs	Conference on Climate Change, Agriculture and Hydropower organized by NGO Forum and Council for Agriculture and Rural Development (CARD), targets included national decision-makers and public (N,P) 3,208 Community representatives from Cambodia provided thumb print to letter to Thai Prime Minister (S)	

(Sources: International Rivers 2013; MRC 2011b, 2011c, 2011d, 2012b; Thai People's Network for Mekong 2011; Save the Mekong 2012; Baran et al. 2011; Trandem 2011; International Rivers 2011; 263 NGOs 2011)

Activities targeted at: (R) regional decision-makers; (N) national decision-makers; (S) stakeholders in potentially affected areas; (P) the general public.

unilaterally terminate the PNPCA process (Higgs 2011). In addition, referring to 'good faith and transparency', which is one of the key principles of the PNPCA (Article 3 PNPCA 2003), the review pointed out that Laos had ignored the concerns raised by other member countries and that it had not conducted a transparent stakeholder consultation process (Higgs 2011). Similarly, the Mekong Legal Network (MLN) wrote a legal briefing note on the Xayaburi PNPCA, which was sent to the members of the MRC Council, JC, Secretariat, and Foreign Ministers of the member countries (Mekong Legal Network 2011: 1). The MLN is a network consisting of lawyers primarily from the Mekong region, aiming to promote the rule of law particularly on cross-border issues, including hydropower dams. It is fostered by Earthrights International, a US-based human rights NGO (Earthrights International 2012). Similarly to the Perkins Coie legal review, the review by the MLN pointed out that Laos' unilateral termination of the PNPCA was a breach of its obligations (Mekong Legal Network 2011).

In its attempt to legitimize the termination of the PNPCA process, the government of Laos commissioned a Finnish consultancy company, Pöyry, to review Laos' compliance with the PNPCA process. The compliance report prepared by Pöyry in August 2011 (referred to as the Pöyry report) reviewed whether the Xayaburi project owner (Xayaburi Power Company) had complied with the MRC's Preliminary Design Guidance for Proposed Hydropower Dam in the Lower Mekong Basin (2009), and whether the Government of Laos and the project owner addressed the concerns raised by other MRC member countries during the prior consultation process (Government of Lao PDR, and Pöyry 2011). It also reviewed points raised in the prior consultation project review report commissioned by the MRC during the initial six-month period of the PNPCA process. The report concluded that the Xayaburi hydropower dam, in principle, follows the MRC design guidelines. However, it also recommended technical improvement to the dam (Government of Lao PDR, and Pöyry 2011: 14).

The Pöyry report was heavily criticized by various actors. In November 2011 *The Cambodia Daily*, an English newspaper circulated in Cambodia, reported that the CNMC Secretary General Te Navuth expressed concern over the gap in knowledge of the Pöyry report, claiming that Laos 'is not looking far from the dam, just a few kilometres away in their own country' (Chen and Bopha 2011). The same newspaper article also provided a comment by Nao Thuok, director general of the Fisheries Administration, claiming that the Xayaburi dam would directly affect six million people whose livelihoods depend on the river (Chen and Bopha 2011). On the NGO side, a review by IR pointed out a number of faults in the Pöyry report, including its failure to respond to other MRC member states' concerns and to meet the MRC's design requirements (Herbertson 2011b). The IR report also criticized the contradictions and inadequacy of the science base used in the Pöyry report, which claimed that the Xayaburi dam met the MRC requirements, while at the same time it also recommended conducting more than 40 studies before the project was in compliance (Herbertson 2011a, 2011b). The WWF also published a review report on the Pöyry report, particularly from a fisheries perspective, pointing out the weakness of the proposed fish passes and the EIA (WWF 2011).

In response to the criticisms against the Pöyry report, the Lao government commissioned a French engineering company, Compagnie Nationale du Rhône (CNR), to conduct a peer review of the Pöyry report (EB *et al.* 2012). The CNR report was finalized in March 2012, and concluded that the Xayaburi project can comply with the MRC guidelines and other best practices on hydropower dams globally if suggestions for improvements are taken into account (EB *et al.* 2012: 20). The report was again criticized by civil society groups. IR pointed out the absence of an assessment of transboundary impacts, as well as the report's assumption concerning sediment transports (International Rivers 2012). WWF also criticized the CNR report for its failure to evaluate fish and fisheries (Worrell 2012).

In December 2011, the ministers of four MRC member countries met at the eighteenth MRC Council meeting (MRC 2012b). The participants discussed the Xayaburi hydropower dam during this meeting and agreed to approach the Japanese government for support in conducting the impact study of the mainstream hydropower projects (MRC 2012b: 3). This solution had been verbally agreed among the Prime Ministers at the third Mekong–Japan Summit held in November 2011 (MRC 2012b: 3). The official meeting minutes do not clearly indicate an extension of the PNPCA period or deferral of the Xayaburi dam construction until the impact study is completed. However, as reported in the regional media, it appeared to be widely understood as a deferral of the plan until the impact studies were completed (*Bangkok Post* 2011b; Lipes 2011) by all except one country, Laos. According to one interviewee, after the press conference of the MRC Council meeting, Laos read a statement indicating that it considered the prior consultation process of the 1995 Mekong Agreement was completed. This surprised many of those who were present at the scene (Lao PDR 2011; V16 2011).

Post–MRC Council meeting

The MRC Council meeting in December 2011 concluded that there was a need for further study before making a final decision on the Xayaburi dam, and the MRC initiated the process of designing further studies on the impacts of the mainstream dams. While the process of agreeing on the ToR of the studies among the four countries was delayed, the Vietnamese government initiated the process of undertaking its own study, focusing on the hydropower dam's impact on the Mekong Delta. This was to be conducted in collaboration with the Cambodian government (R2 2012). The contract to conduct the study was later signed in June 2013 with a Danish consultancy company and the VNMC (Huong 2013). Concurrently, Laos continued to move the Xayaburi dam plan forward, and in April 2012 the construction contract was signed between

Table 7.3 Key activities January–August 2012

Time	Key events	Key reactions by NGOs in the Mekong region	RCC activity	VRN activity
Jan 2012			Fifty monks in Kampong Cham province attended training on Xayaburi, conducted by RCC (S)	
Feb 2012			Pray for river event organized by 3S Rivers Protection Network (3SPN) calling for cancellation of Xayaburi dam and ten-year study (S)	
Mar 2012	CNR report published	IR and WWF separately criticized the report (R)		
Apr 2012	CH. Karnchang informed Stock Exchange of Thailand that it had signed a construction contract to build Xayaburi dam (17 Apr)	Thai community groups protested in front of CH. Karnchang and MRC meetings (R)		
Apr–Jun 2012	Government of Laos made inconsistent statements regarding Xayaburi dam construction	STM Coalition sent letters to MRC and ministers of Thailand, Cambodia, and Vietnam requesting clarification of PNPCA status (R)	Some RCC members signed letter from STM Coalition (R)	VRN signed letter by STM Coalition (R)
		IR conducted site visit to Xayaburi dam (Jun), filmed construction, and broadcast footage via media including BBC (Jul) (P, R)		

Time	Key events	Key reactions by NGOs in the Mekong region	RCC activity	VRN activity
Jun 2012		Coalition of NGOs led by Siemmenpu foundation (Finnish NGO) filed complaint to Finnish OECD focal point regarding Pöyry (R)	RCC joined complaint to Pöyry (R) Peace Walk organized by RCC in Kampong Cham, during which 186 communities, monks, students, and CSOs signed petition to stop dam (S) Two awareness-raising events held in Kratie province (S)	VRN joined complaint to Pöyry (R)
Jul 2012	Laos invited development partners and NGO representatives to visit Xayaburi dam site, officially admitting ongoing construction		RCC wrote to Prime Ministers of Thailand and Laos requesting stop to Xayaburi dam construction and cancellation of power purchase agreement (R)	VRN organized workshop for provincial governments and farmers in Mekong Delta (S)
Aug 2012		A group of Thai villagers from riparian villages of the Mekong River filed court case against EGAT power purchase agreement (R)	RCC posted opinion article in *Bangkok Post* pleading for stop to Xayaburi dam (R)	VRN organized workshop with VUSTA – 'The Mekong and hydropower' – targeting National Assembly members, scientists, journalists and NGOs (N, P)
Nov 2012	Laos officially launched Xayaburi dam construction			
Jan 2013	During MRC Council meeting Cambodia and Vietnam requested further impact assessments			

(Sources: International Rivers 2013; MRC 2011b, 2011c, 2011d, 2012b; Thai People's Network for Mekong 2011; Save the Mekong 2012; Baran *et al.* 2011; Trandem 2011; International Rivers 2011; 263 NGOs 2011; Vandenbrink 2012; Deetes 2012)

Activities targeted at: (R) regional decision-makers; (N) national decision-makers; (S) stakeholders in potentially affected areas; (P) the general public.

CH. Karnchang public company and the Xayaburi Power Company (CH. Karnchang 2012).

These actions led to further public criticism against the Xayaburi dam. In Thailand, villagers from eight provinces along the Mekong River rallied in front of CH. Karnchang's headquarters in Bangkok, demanding to stop the Xayaburi dam construction (Wannamontha 2012; Wiriyapong 2012). Further protests were conducted outside the 'Mekong2Rio' conference in Phuket in May 2012. This conference gathered participants from various river basins across the world with the aim of promoting the sustainable use of water as a key agenda item for the Rio+20 Summit in June 2012 (MRC 2012a; Ganjanakhundee 2012b).

The governments of Cambodia and Vietnam also responded to the Lao government's actions associated with the dam. The Cambodian Minister of Water Resources and Meteorology, Lim Kean Hor, sent a letter to Laos in April 2012 requesting a halt to the construction of the Xayaburi dam and requesting detailed information about the impact on the Mekong River from the dam project (Narin and Chen 2012). Newspapers reported that the permanent vice-chairman of the CNMC had warned Laos of potentially bringing the case before the international court (Vandenbrink 2012a). He did not indicate any specific courts; however, these types of dispute are typically handled by the International Court of Justice (Vandenbrink 2012a). The VNMC also criticized the construction plans, condemning CH. Karnchang's actions after the company notified signing the construction agreement on the Xayaburi dam in April 2012 and then notified the Thai Stock Exchange (Phong 2012; CH. Karnchang 2012). At this stage, Vietnam condemned CH. Karnchang but not the Lao government, claiming that the company violated the joint agreement made by the MRC member countries, an agreement that included Laos (Phong 2012).

The international donor community also raised concerns over the Xayaburi hydropower dam. The Xayaburi dam PNPCA process was raised at several informal donor meetings between development partners and the MRC members (MRC 2011e). The development partners urged the MRC members to clarify the status of the PNPCA and offered financial or technical assistance in moving any required assessment forward (Development Partners to the MRC 2012; MRC 2011a, 2011e). Some donors directly urged Laos to follow the consensus reached at the MRC meetings. For example, US Secretary of State Hillary Clinton met with Lao Prime Minister Thongsing Thammavong during her visit to the region in July 2012, urging the Lao Prime Minister to put the Xayaburi dam on hold until impact studies were conducted (AFP 2012; OOSKAnews Correspondent 2012).

There were inconsistencies in the Lao government's reactions to these criticisms. On one hand, Laos appeared to continue construction of the Xayaburi dam. Viraphonh Viravong, the Lao Vice Minister of Energy and Mines, commented that 'we will address all reasonable concerns in order to make this Xayaburi dam a transparent dam and a role model for other dams in the mainstream of the Mekong River' (Ganjanakhundee 2012a; Hunt 2012). On the other hand, Viravong was also reported as commenting that Laos would not make a final decision on the Xayaburi dam construction without the approval of the international community and the Lower Mekong riparian countries (Ganjanakhundee 2012a). Another senior official from the Lao Ministry of Foreign Affairs told Radio Free Asia, one of the international media outlets in the region, that construction was discontinued and postponed, stating that Laos would comply with the 1995 Mekong Agreement (Vandenbrink 2012e). Subsequently, in June 2012, during the sidelines of the eighteenth ASEAN summit in Jakarta, Vietnamese Prime Minister Nguyen Tan Dung praised Lao Prime Minister Thoongsing Thammavong for Laos' decision to halt the Xayaburi dam construction (*Thanh Nien News* 2011).

However, in reality, construction associated with the Xayaburi dam was continuing, and in June 2012 IR visited the Xayaburi dam site to film the ongoing construction. The footage was broadcast globally through BBC World News in July 2012, which again attracted criticism against Laos (Fisher 2012a). In July 2012, the Vietnamese and Cambodian prime ministers agreed to sign a joint letter to Laos and Thailand requesting more time for a comprehensive impact assessment before a decision was made on the dam (Vandenbrink 2012d; *The Phnom Penh Post* 2012). At the end of July 2012, the government of Laos organized a site visit for development partners, officially revealing the construction of the Xayaburi dam (Fisher 2012b; Ponnudurai 2012).

In another effort to try to stop the construction of the Xayaburi dam, 37 community members living along the Mekong River in Thailand filed a lawsuit at the Administrative Court in Bangkok (Ten 2012; Deetes 2012). The lawsuit was filed against EGAT, the Thai Cabinet, and three other entities, claiming that the Thai government allowed EGAT to sign the power purchase agreement with the Xayaburi Power Company without a proper impact assessment for the project (Vandenbrink 2012c). The Administrative Court of Thailand later claimed that it did not have the jurisdiction to hear this case since the power purchase agreement was binding between EGAT and the Xayaburi Power Company, and the community was not considered as injured persons (Chuang 2013). The decision also indicated that it was not within the capacity of the administrative court to hear a case of compliance with the PNPCA (Chuang 2013).

While the concerns of the public and other riparian countries still remained, Laos officially launched the construction of the Xayaburi hydropower dam in November 2012 (Ministry of Energy and Mines 2012). In its official statement, Laos again reiterated its understanding that Laos complied with the 1995 Mekong Agreement, and that it had improved the design of the dam as a result of taking into consideration other riparian states' concerns (Ministry of Energy and Mines 2012). The official launch ceremony was attended by ambassadors from Cambodia and Vietnam, two downstream countries which had raised concerns over the dam (*Vientiane Times* 2012; Vandenbrink 2012b). However, during the MRC Council meeting held in January 2013, representatives of these two countries again demanded an impact study on the Xayaburi (Vandenbrink 2013). Despite the concerns, no further measures to stop the Xayaburi construction, or measures to resolve the contradiction, were taken. At the time of writing this book, construction continues at the Xayaburi dam site (Wangkiat 2015).

Conclusion

The review of actors and their positions during the Xayaburi PNPCA process provides some insights into the analysis of NGO strategies. First, the position of each government became clear at the end of the initial six months of the PNPCA process. Gaining clarity over the two governments' positions (Vietnam and Cambodia) was positive for both the RCC and the VRN in their subsequent advocacy work with their national decision-makers, as discussed in Chapter 8.

Second, the review identifies the shift in the forum for the states' negotiations over the Xayaburi dam over time. During the first six months of the PNPCA (until the JC meeting in April 2011), the MRC member states' activities and negotiations took place primarily within the MRC. However, many discussions following the initial six-month period took place at various venues outside the MRC, including side discussions during meetings between state representatives at events such as the ASEAN meeting, or in bilateral meetings. This shift in the venue of discussion had implications for the NGOs that aimed to influence the decision-making process.

Finally, and most importantly, the implementation of the PNPCA process of the Xayaburi dam was to a large extent dependent on actors' interpretations of the requirements and ambiguities in the requirements. This situation created difficulties for NGOs in both Vietnam and Cambodia, as well as for the negotiating governments themselves, and hindered a clear conclusion as to the way forward in the process.

One of the ambiguities was in determining when the prior consultation process could be considered 'complete'. The 1995 Mekong Agreement requires prior consultation to aim at arriving at an agreement by the JC (Article 5B *Mekong Agreement* 1995). This requirement does not make clear whether the consultation is considered complete when the member states aim to arrive at an agreement, but do not actually reach such an agreement; or whether it is considered complete only if the member state countries have come to an agreement before the consultation process ends. The lack of clarity on how to extend the duration of prior consultation was another associated problem. The PNPCA indicates that 'the timeframe for the Prior Consultation shall be six months from the date of receiving documents on Prior Consultation' (5.5.1 PNPCA 2003). However, it also indicates that 'if necessary, an extended period shall be permitted by the decision of the MRC JC' (5.5.2 *PNPCA* 2003).

In the case of the Xayaburi dam, at the end of six months from the time the PNPCA process commenced, MRC JC members were unable to unanimously conclude the consultation process, and therefore agreed that the decision on the prior consultation process of the Xayaburi dam be tabled at the ministerial level (MRC 2011b). The official minutes from this JC meeting are not publicly available, and the exact wording agreed during this meeting is not clear. However, two months after the MRC JC meeting, the government of Laos expressed its opinion that it had provided an opportunity for each MRC member country to evaluate, discuss, and comment on the Xayaburi project, and that it had taken all the concerns from the member countries into consideration. Therefore it considered the Xayaburi prior consultation process was complete (Phomsoupha 2011), a decision that attracted the criticism of other riparian countries and other actors such as MRC donors and NGOs (MRC 2011a). Even the MRC Secretariat was unable to provide a clear answer on the status of the PNPCA process. In response to a letter from a regional NGO coalition received 20 months after the PNPCA process started, the CEO of the MRC indicated that 'due to differing views of the four countries no consensus could be reached on whether or not the consultation should be considered complete' (Guttman 2012).

This situation revealed another associated problem for the Mekong Agreement, namely that it does not provide a detailed dispute resolution mechanism. The 1995 Mekong Agreement discusses how to address the differences and disputes (Chapter V *Mekong Agreement* 1995) and suggests resolution through discussions within the MRC (Article 34 *Mekong Agreement* 1995). In case the difference or dispute cannot be resolved through the MRC, the Agreement suggests that countries resolve the issues through diplomatic channels, and in the last resort, 'seek assistance to mediation through an entity or party mutually agreed upon prior to proceeding to the conflict resolution mechanism according to the principles of international law' (Article 35 *Mekong Agreement* 1995). This conflict resolution mechanism under the 1995 Mekong Agreement is weak compared with other agreements in the area of international waters. For example, the United Nations Convention on the Law of the Non-Navigational Uses of International Watercourses (1997 UN Watercourses Convention) provides a detailed procedure which parties can consider taking in case of dispute, such as submission of the dispute to the International Court of Justice, and/or the establishment of fact-finding commissions (Article 33 *UN Watercourses Convention* 1997). This somewhat 'soft' nature of dispute resolution mechanism under the 1995 Mekong Agreement may reflect the political culture in Southeast Asia, described as the 'ASEAN Way', which incorporates social characteristics of Southeast Asian culture such as 'conflict avoidance and harmony, consensual group behaviour, personal relationships taking precedence over other relationships in politics or business, and indirectness and circumlocution in communication' (Hirsch *et al.* 2006: 76).

There was also ambiguity about how to ensure public participation and transparency of the process. While public participation is not a mandatory requirement under the PNPCA process, the PNPCA specifies that one of the roles and functions of the National Mekong Committees is 'to facilitate any consultation, presentations, evaluation and site visit as requested by the MRC

JC for the proposed use' (Section 5.3.1 d. *PNPCA* 2003). During the PNPCA process, the PNPCA JC working group agreed that stakeholder consultation should be considered as a national matter (MRC 2011c: 2). Organizing public consultation was therefore left to each NMC's discretion within their jurisdiction, creating differences in approach. The criticism by the NGOs that there was a lack of sufficient time for public consultation and participation reflects the fact that existing MRC policies and guidelines on stakeholder participation (discussed in Chapter 6) that specifically suggest these points were not fully applied by the member states.

While 'transparency' is one of the key principles under which the PNPCA process should be governed (Section 3 *PNPCA* 2003), it does not specifically provide a mandate for states to act transparently on every decision or action. One of the points about the process of the Xayaburi PNPCA that NGOs criticized was the lack of transparency and information disclosure (Save the Mekong 2011). For instance, although all the technical documents later became publicly available through the MRC's websites (MRC 2011d), when public consultation took place in early 2011 only the feasibility study was available to the public, not the EIA (MRC 2011c: 2). The MRC prior consultation report indicates that the choice of disclosure of certain documents was the decision of the PNPCA Joint Committee Working Group, which 'considered that any submitted documents could only be released and/or disseminated beyond the MRC framework with the official permission of the submitting country, in this case the Lao PDR' (MRC 2011c: 2). This disclosure status indicates a weakness in the existing MRC framework. The MRC disclosure policy indicates that the project-related EIA should be open to the public after the report is made public under prevailing national regulations (Section 4 MRC 2007). In other words, if some member countries have weak domestic laws associated with access to information, it hinders transparency at the regional scale.

These ambiguities of the Mekong Agreement and the PNPCA created a situation in which NGOs in the region demanded more clarity and openness in the states' discussions. The situation illustrates the potential importance of formal rules such as the MRC Agreement in shaping the advocacy strategies of NGOs and civil society coalitions.

Tables 7.1–7.3 highlight the activities of the RCC and VRN during this two-year period. The activities are categorized into four types according to the target audience. This categorization of advocacy strategies provides the structure for the analysis in the rest of this book. Chapters 8–11 compare the strategies adopted by each coalition, and analyse how rules and norms shaped their strategies. The analysis compares the Vietnamese and Cambodian strategies targeting the same audiences: Chapter 8 strategies targeting regional decision-makers; Chapter 9 those targeting national decision-makers; Chapter 10 those targeting stakeholders who face potential impact from the Xayaburi dam, and Chapter 11 targeting the general public. Applying the framework for understanding advocacy strategies in Figure 2.1, Figure 7.1 illustrates these categories of strategy and the overall structure of the book.

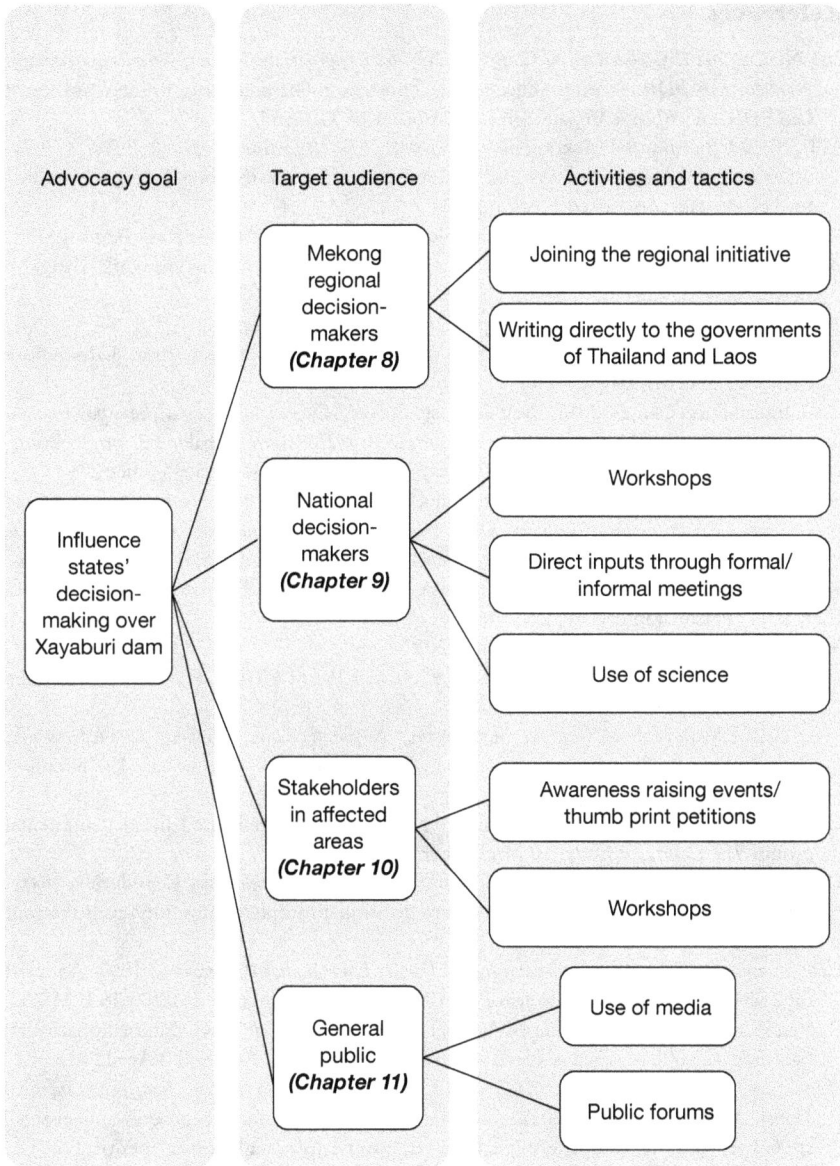

Figure 7.1 Structure of chapters discussing advocacy strategies

References

263 NGOs. 2011. *Global Call to Cancel the Xayaburi Dam on the Mekong River mainstream in Northern Lao PDR*. A letter addressed to Thongsing Thammavong, Prime Minister of Lao PDR and Abhisit Vejjajiva, Prime Minister of Thailand.

AFP. 2012. Clinton urges Mekong nations to avoid US dam mistakes. *Bangkok Post*, 14 July. www.bangkokpost.com/news/asia/302366/clinton-urges-mekong-nations-to-avoid-us-dam-mistakes. Accessed 5 April 2013.

Agreement on the Cooperation for the Sustainable Development of the Mekong River Basin. 1995.

Bangkok Post. 2011a. Laos defends jumping gun on Xayaburi construction work. Bangkok Post, 8 May.

—— 2011b. Mekong panel stalls dam plan. *Bangkok Post*, 9 December.

—— 2011c. Xayaburi dam work begins on sly: Thai construction giant, Laos ignore Mekong concerns. *Bangkok Post*, 17 April.

Bank Information Center. 2011. Press Release: *263 NGOs globally call on Mekong governments to cancel plans to build Xayaburi dam*. www.bicusa.org/263-ngos-globally-call-on-mekong-governments-to-cancel-plans-to-build-xayaburi-dam. Accessed 5 September 2013.

Baran, Eric, Michel Larinier, Guy Ziv, and Gerd Marmulla. 2011. *Review of the Fish and Fisheries Aspects in the Feasibility Study and the Environmental Impact Assessment of the Proposed Xayaburi Dam on the Mekong Mainstream*. WWF Greater Mekong. http://assets.panda.org/downloads/wwf_xayaburi_dam_review310311.pdf. Accessed 19 April 2011.

C29. 2011. Personal interview, 20 November 2011.

—— 2012. Personal interview, 10 August 2012.

CH. Karnchang. 2011. *Building a Better Life: Annual Report 2010*. CH. Karnchang Public Company Limited.

—— 2012. *Notification of Contract Signing with Xayaburi Power Company Limited*. www.ch-karnchang.co.th/news_activities_detail_en.php?nid=474. Accessed 19 October 2012; no longer available online.

Chen, Dene-Hern, and Phorn Bopha. 2011. Lao dam report blasted by environmental group. *The Cambodia Daily*, 10 November.

Chuang, Evelyn. 2013. *Xayaburi Dam Court Decision Prompts Community Consultation*. www.earthrights.org/blog/xayaburi-dam-court-decision-prompts-community-consultation. Accessed 2 April 2013.

Convention on the Law of the Non-navigational Uses of International Watercourses. 1997. Adopted on 21 May 1997, entered into force on 17 August 2014. Reprinted (1997) 36 ILM 700.

Deetes, Pianporn. 2012. *Thai Villagers File Lawsuit on Xayaburi Dam*. www.internationalrivers.org/blogs/254/thai-villagers-file-lawsuit-on-xayaburi-dam. Accessed 2 April 2013.

Development Partners to the MRC. 2012. *Joint Development Partner Statement: Informal Donor Meeting*, 29 June 2012. Vientiane. www.mrcmekong.org/news-and-events/speeches/joint-development-partner-statement-informal-donor-meeting-2012-vientiane-lao-pdr. Accessed 3 May 2014.

Earthrights International. 2012. *Mekong Legal Network*. www.earthrights.org/legal/mekong-legal-network. Accessed 8 April 2013.

EB, EPP, and MEM. 2012. *Xayaburi Hydroelectric Power Project: Peer Review of the Compliance Report made by Pöyry*. Final Report. Department of Energy Business (EB), Energy Policy and Planning Department (EPP), and Ministry of Energy and Mines-Lao PDR. www.poweringprogress.org/download/Reports/2012/April/Final-report-V1.pdf. Accessed 6 September 2012.

Fisher, Jonah. 2012a. Laos' work on the Mekong river draws criticism. *BBC World News*, 4 July. www.bbc.co.uk/news/world-18700473. Accessed 27 March 2013.

—— 2012b. Team visits controversial Laos dam. *BBC News Asia*, 26 July. www.bbc.co.uk/ news/world-asia-18993032. Accessed 28 March 2013.

Ganjanakhundee, Supalak. 2012a. Laos: no work on Xayaburi dam until green concerns solved. *The Nation*, 4 May. www.nationmultimedia.com/politics/Laos-no-work-on-Xayaburi-dam-until-green-concerns--30181251.html. Accessed 27 March 2013.

—— 2012b. Residents demand answers on Xayaburi. *The Nation*, 2 May. www.thaivisa. com/forum/topic/552013-residents-demand-answers-on-xayaburi-dam. Accessed 3 May 2012.

Government of Lao PDR, and Pöyry. 2011. *Compliance Report: Xayaburi Hydroelectric Power Project. Run-of-River Plant*. www.poweringprogress.org/download/Reports/2012/July/ Compliance%20Report%20Xayaburi%20Main%20Final.pdf. Accessed 7 June 2015.

Guttman, Hans. 2012. *Reply to your request for clarifications on the Prior Consultation for the proposed Xayaburi project*. A letter to Save the Mekong Coalition members, 14 May. www.savethemekong.org/admin_controls/js/tiny_mce/plugins/imagemanager/files/ May142012.pdf. Accessed 7 June 2015.

HELVETAS Laos. 2011. Letter to MRC Secretariat, 15 February 2011. www.mrcmekong. org/pnpca/petitions/Helvetas-letter-15Feb11.pdf. Accessed 11 April 2011; no longer available online.

Herbertson, Kirk. 2011a. *Review of the Pöyry report on the Xayabguri Dam*. www. internationalrivers.org/resources/review-of-the-p%C3%B6yry-report-on-the-xayaburi-dam-3929. Accessed 26 March 2013.

—— 2011b. *Sidestepping Science: Review of the Pöyry Report on the Xayaburi Dam*. International Rivers. www.internationalrivers.org/files/attached-files/intl_rivers_analysis_of_poyry_ xayaburi_report_nov_2011.pdf. Accessed 29 May 2012.

Higgs, Stephen J. 2011. *Re: PNPCA Process for Xayaburi Dam*. A letter to Aviva Imhof, International Rivers and Lewis Gordon, Environmental Defender Law Center, 5 July. Portland, OR: Perkins Coie. www.internationalrivers.org/files/attached-files/xayabur ipnpcaprocess.pdf. Accessed 13 July 2015.

Hirsch, Philip, Kurt Mørck Jensen, Ben Boer, Naomi Carrard, Stephen FitzGerald, and Rosemary Lyster. 2006. *National Interests and Transboundary Water Governance in the Mekong*. http://sydney.edu.au/mekong/documents/mekwatgov_mainreport.pdf. Accessed 23 July 2011.

Hunt, Luke. 2012. Laos: Xayaburi dam halted. Really. The Diplomat, 5 May. http:// thediplomat.com/asean-beat/2012/05/05/laos-xayaburi-dam-on-hold-really. Accessed 28 March 2013.

Huong, Le Ti. 2013. *Viet Nam lands a contract with Danish Consultant for a study on impacts of mainstream hydropower on the Mekong River*. http://vnmc.gov.vn/newsdetail/345/viet-nam-lands-a-contract-with-danish-consultant-for-a-study-on-impacts-of-mainstream-hydropower-on-the-mekong-river.aspx. Accessed 21 June 2013.

ICEM. 2010a. *MRC Strategic Environmental Assessment (SEA) of Hydropower on the Mekong mainstream: Summary of the Final Report*. Hanoi: Viet Nam. www.mrcmekong.org/assets/ Publications/Consultations/SEA-Hydropower/SEA-FR-summary-13oct.pdf. Accessed 7 June 2015.

—— 2010b. *MRC Strategic Environmental Assessment (SEA) of the Hydropower on the Mekong Mainstream*. Mekong River Commission. Hanoi: Vietnam. www.mrcmekong.org/ assets/Publications/Consultations/SEA-Hydropower/SEA-Main-Final-Report.pdf. Accessed 3 May 2014.

International Rivers. 2011. *263 NGOs Call on Mekong Governments to Cancel Plans for Xayaburi Dam.* www.internationalrivers.org/resources/263-ngos-call-on-mekong-governments-to-cancel-plans-for-xayaburi-dam-3728. Accessed 7 June 2015.

—— 2012. *Comments on CNR's report for the Government of Laos on the Xayaburi Dam.* www.internationalrivers.org/review-of-cnr%E2%80%99s-report-for-laos-on-the-xayaburi-dam-june-2012. Accessed 6 September 2012.

—— 2013. *Xayaburi Dam: Timeline of Events.* www.internationalrivers.org/files/attached-files/xayaburi_dam_timeline_of_events_feb._2013_0.pdf. Accessed 28 March 2013.

Kingdom of Cambodia. 2011. *Mekong River Commission Procedures for Notification, Prior Consultation and Agreement: Form/Format for Reply to Prior Consultation.* Cambodia National Mekong Committee. www.mrcmekong.org/assets/Consultations/2010-Xayaburi/Cambodia-Reply-Form.pdf. Accessed 25 November 2011.

Lao PDR. 2011. Statement read by Lao PDR after the MRC Council meeting on 8 December 2012 (unofficial document obtained from interviewee).

Lipes, Joshua. 2011. Dam deferred pending impact study. *Radio Free Asia*, 12 April. www.rfa.org/english/news/laos/dam-12082011151831.html?searchterm=Xayaburi. Accessed 27 March 2013.

Mekong Legal Network. 2011. *Briefing Note on Duties of Notification, Prior Consultation, and Assessment Arising From International Law in Relation to the Xayaburi Dam Project in Northern Lao PDR.* http://w4pn.org/index.php/w4pn-resources-download/doc_view/54-memorandum-on-legal-aspects-of-the-power-purchase-agreement-for-the-xayaburi-hydropower-project.raw?tmpl=component. Accessed 7 June 2015.

MRC. 2007. *MRC Policy on Disclosure of Data, Information and Knowledge. Mekong River Commission.* www.mrcmekong.org/assets/Publications/policies/Communication-Strategy-n-Disclosure-Policy.pdf. Accessed 3 May 2014.

—— 2011a. *Joint Development Partner Statement: Donor Consultative Group – Eighteenth MRC Council Meeting, Joint Meeting with the MRC Donor Consultative Group.* Mekong River Commission. www.mrcmekong.org/news-and-events/speeches/joint-development-partner-statement-donor-consultative-group-18th-mrc-council-meeting-joint-meeting-with-the-mrc-donor-consultative-group. Accessed 7 June 2015.

—— 2011b. Media Release: *Lower Mekong Countries Take Prior Consultation on Xayaburi Project to Ministerial Level.* Mekong River Commission. www.mrcmekong.org/news-and-events/news/lower-mekong-countries-take-prior-consultation-on-xayaburi-project-to-ministerial-level. Accessed 7 June 2015.

—— 2011c. *Prior Consultation Project Review Report: Volume 2 Stakeholder Consultations Related to the Proposed Xayaburi Dam Project.* Mekong River Commission. www.mrcmekong.org/assets/Consultations/2010-Xayaburi/2011-03-24-Report-on-Stakeholder-Consultation-on-Xayaburi.pdf. Accessed 7 June 2015.

—— 2011d. *Proposed Xayaburi Hydropower Project MRC's Prior Consultation Process* (2011 version of website). Mekong River Commission. www.mrcmekong.org/pnpca/PNPCA-technical-process.htm. Accessed 11 April 2011; no longer available online. Updated version at www.mrcmekong.org/news-and-events/consultations/xayaburi-hydropower-project-prior-consultation-process. Accessed 7 June 2015.

—— 2011e. Report: *Informal Donor Meeting*, 23–24 June 2011. Phnom Penh: Mekong River Commission. www.mrcmekong.org/assets/Publications/governance/Minutes-of-IDM2011-final.pdf. Accessed 31 October 2011.

—— 2012a. *Mekong2Rio.* Mekong River Commission. www.mrcmekong.org/news-and-events/events/mekong2rio/?url=/mekong2rio. Accessed 4 April 2013.

—— 2012b. *Minutes of the Eighteenth Meeting of the MRC Council*, 8 December 2011. Siem Reap, Cambodia: Mekong River Commission. www.mrcmekong.org/assets/ Publications/governance/Minutes-of-the-18th-Council.pdf. Accessed 7 June 2015.

MRC Secretariat. 2011. *Proposed Xayaburi Dam Project-Mekong River: Prior Consultation Project Review Report*. Mekong River Commission Secretariat. www.mrcmekong.org/ assets/Publications/Reports/PC-Proj-Review-Report-Xaiyaburi-24-3-11.pdf. Accessed 25 October 2011.

Ministry of Energy and Mines. 2012. *Statement of the Government of Lao PDR on Xayaburi Hydropower Project*. www.poweringprogress.org/index.php?option=com_content& view=article&id=947:19th-november-2012-statement-of-the-government-of-lao-pdr-on-xayaburi-hydropower-project&catid=85:press-releases&Itemid=50. Accessed 28 March 2013.

Narin, Khuon, and Dene-Hern Chen. 2012. Cambodia asks Laos to halt all activity on Mekong dam. *The Cambodia Daily*, 2 May.

NGO Forum on Cambodia. 2011. *Thank you letter for your kind collaboration with Rivers Coalition in Cambodia and civil society concerns on the proposed Xayaburi dam*. A letter to H.E. Lim Kean Hor, Chairman of Cambodia National Mekong Committee, 23 March 2011.

OOSKAnews Correspondent. 2012. Clinton urges further studies on Xayaburi dam construction. *OOSKAnews*, 11 July. www.ooskanews.com/daily-water-briefing/clinton-urges-further-studies-xayaburi-dam-construction_23336. Accessed 5 April 2013.

Phomsoupha, Xaypaseuth, 2011. *Re: Xayaburi Hydroelectric Power Project*. A letter to Trivisvavet, Thanswat (Xayaburi Power Company Limited), 8 June, from the Director General. Ministry of Energy and Mines, Department of Energy Promotion and Development, Lao PDR.

Phong, Nguyen. 2012. MRC Vietnam condemns Thai company's contract to build Xayaburi dam. *Thanh Nien News*, 24 April. www.thanhniennews.com/politics/mrc-vietnam-condemns-thai-companys-contract-to-build-xayaburi-dam-7641.html. Accessed 7 June 2015.

Ponnudurai, Parameswaran. 2012. Lao dam PR blitz backfires. *Radio Free Asia*, 26 July. www.rfa.org/english/commentaries/east-asia-beat/xayaburi-07262012050806. html?searchterm=Xayaburi. Accessed 28 March 2013.

Procedures for Notification, Prior Consultation and Agreement (PNPCA). 2003. Approved by the 10th meeting of the MRC Council on 30 November 2003.

R2. 2011. Personal interview, 16 November 2011.

—— 2012. Personal interview, 30 June 2012.

R6. 2012. Personal interview, 4 July 2012.

Roeun, Van, and Clancy McGillian. 2010. Groups, fishermen voice concern over dam. *The Cambodia Daily*, 24 September.

Save the Mekong. 2010. *Subject: Call for Halt to the PNPCA Process and Cancellation of Xayaboury Dam*. A letter to Mr Jeremy Bird, CEO, Mekong River Commission, 13 October.

—— 2011. *Subject: Request for MRC Council to halt the current PNPCA process on the Xayaburi Dam, to endorse the MRC's SEA report's findings, and commit to evaluate all options for meeting the Mekong region's water and energy needs through a credible and objective public process*. A letter to H.E. Mr Lim Kean Hor: Minister of Water Resources and Meteorology et al., 25 January.

—— 2012. *Subject: Request for Clarifications on the Prior Consultation for the Xayaburi Dam*. A letter to Mr Hans Guttman, CEO, Mekong River Commission, 20 April.

Socialist Republic of Viet Nam. 2011. *Mekong River Commission Procedures for Notification, Prior Consultation and Agreement Form for Reply to Prior Consultation.* The Viet Nam National Mekong Committee. www.mrcmekong.org/assets/Consultations/2010-Xayaburi/Viet-Nam-Reply-Form.pdf. Accessed 25 October 2011.

Ten, Daniel. 2012. Thai lawsuit threatens to derail laos plans for Mekong River dam. *Bloomberg News*, 7 August. www.bloomberg.com/news/articles/2012-08-07/thai-lawsuit-threatens-to-derail-Laos-plans-for-mekong-river-dam. Accessed 7 June 2015.

Thai People's Network for Mekong. 2011. Open letter. *A letter to Mr. Jeremy Bird, CEO of the Mekong River Commission*, 19 January.

Thanh Nien News. 2011. Vietnam hails Laos for suspending Xayaburi dam. *Thanh Nien News*, 29 June. www.thanhniennews.com/politics/vietnam-hails-laos-for-suspending-xayaburi-dam-12416.html. Accessed 7 July 2012.

The Phnom Penh Post. 2012. Cambodia, Vietnam united on Xayaburi. *The Phnom Penh Post*, 5 July 2012. www.eco-business.com/news/cambodia-vietnam-united-on-xayaburi/. Accessed 7 July 2012.

Trandem, Ame. 2011. *Milestones of Concern: A Timeline of Concerns Expressed over the Proposed Xayaburi Dam.* September 2008 to July 2011. (Document obtained from the author.)

V16. 2011. Personal interview, 23 November 2011.

Vandenbrink, Rachel. 2012a. Cambodia warns Laos over Mekong dam. *Radio Free Asia*, 19 April. www.rfa.org/english/news/cambodia/dam-04192012143244.html?searchterm=xayaburi. Accessed 4 June 2012.

—— 2012b. Ground broken on Xayaburi. *Radio Free Asia*, 7 November. www.rfa.org/english/news/laos/xayaburi-11072012163416.html?searchterm=Xayaburi. Accessed 2 April 2013.

—— 2012c. Thai villagers sue over dam. *Radio Free Asia*, 7 August. www.rfa.org/english/news/laos/xayaburi-08072012171723.html. Accessed 2 April 2013.

—— 2012d. Vietnam joins Cambodia on Xayaburi opposition. *Radio Free Asia*, 6 July. www.rfa.org/english/news/laos/xayaburi-07062012163933.html. Accessed 11 July 2012.

—— 2012e. Xayaburi dam construction suspended. *Radio Free Asia*, 9 May. www.rfa.org/english/news/laos/xayaburi-05092012154022.html?searchterm=xayaburi. Accessed 4 June 2012.

—— 2013. Xayaburi concerns mount. *Radio Free Asia*, 18 January. www.rfa.org/english/news/laos/xayaburi-01182013191811.html?searchterm=Xayaburi. Accessed 2 April 2013.

Vientiane Times. 2012. Ground broken for Xayaburi. *Vientiane Times*, 12 November.

Viet Nam News. 2011. Prime Minister Dung holds talks with Cambodian leader Hun Sen. *Viet Nam News*, 25 April. http://vietnamnews.vnagency.com.vn/Politics-Laws/210668/Prime-Minister-Dung-holds-talks-with-Cambodian-leader-Hun-Sen.html. Accessed 20 May 2012.

Vietnam Plus. 2011. PM Dung meets Lao and Thai counterparts. Vietnam Plus, 7 May. http://en.vietnamplus.vn/Home/PM-Dung-meets-Lao-and-Thai-counterparts/20115/18092.vnplus. Accessed 11 July 2012.

Vrieze, Paul, and Phorn Bopha. 2011. Dam delay welcomed, but might not deter Laos. *The Cambodia Daily*, 21 April.

Wangkiat, Paritta. 2015. Trouble swells on the Mekong. *Bangkok Post*, 16 February. www.bangkokpost.com/news/general/475742/trouble-swells-on-the-mekong. Accessed 25 February 2015.

Wannamontha, Thiti. 2012. Xayaburi dam protest. *Bangkok Post*, 24 April. www.bangkokpost.com/multimedia/photo/290238/xayaburi-dam-protest. Accessed 4 June 2012.

Wiriyapong, Nareerat. 2012. Mekong plans anger riverside communities: villagers protest against Xayaburi dam in capital. *Bangkok Post*, 25 April. www.bangkokpost.com/news/local/290301/mekong-plans-anger-riverside-communities. Accessed 4 June 2012.

Worrell, Shane. 2012. Xayaburi study questioned. *The Phnom Penh Post*, 21 May. www.phnompenhpost.com/index.php/2012052156277/National-news/xayaburi-study-questioned.html. Accessed 4 June 2012.

WWF. 2011. *Critical Review of the Pöyry Compliance Report about the Xayaburi Dam and the MRC Design Guidance*. Worldwide Fund for Nature. http://awsassets.panda.org/downloads/review_of_fisheries_aspects_in_the_poyry_report.pdf. Accessed 26 March 2013.

8 Strategies targeting the Mekong regional decision-makers

Introduction

Advocacy attempting to influence international decision-making processes is not a simple task, particularly when there are a large number of states involved. For civil society groups operating at local and national levels, their entry points for influencing a group of international decision-makers may be limited. In order to engage with the international level of actors, civil society actors often formulate transnational networks, and attempt advocacy from different spheres. Transnational networks can play important roles in the advocacy strategies of national NGOs that are part of riparian nations such as those along the Mekong. Another entry point for civil society actors to hold a dialogue with international decision-makers is through a formal process or platform that allows civil society to communicate directly with international decision-makers.

In addressing the decision-makers at the Mekong regional level, both the Rivers Coalition in Cambodia (RCC) and the Vietnam Rivers Network (VRN) worked primarily within an international network of activists, taking part in a transnational advocacy network (TAN). Referring to the concept of TANs, which was introduced in Chapter 2, this chapter discusses the strategies of the RCC and VRN in targeting Mekong regional decision-makers. Following this analysis, the chapter highlights the recurring themes which were identified through analysis of the interview data. This provides an entry point to the analysis of how rules, norms, and actors interacted to create these strategies.

Strategies adopted by the RCC and VRN

During the Xayaburi Procedures for Notification, Prior Consultation and Agreement (PNPCA), both the VRN and RCC took part in activities coordinated by the Save the Mekong (STM) Coalition, a TAN with its focus in the Mekong region, specifically on the hydropower dams on the mainstream of the Mekong River (Save the Mekong 2009). During the Xayaburi PNPCA period, the STM Coalition sent letters addressed to decision-makers and key actors at the Mekong regional level, including Mekong River Commission

(MRC) Council members from the four member countries, the Prime Ministers of Thailand and Laos, and the CEO of the MRC (Save the Mekong 2010, 2011, 2012a, 2012b). As members of the STM Coalition, both the RCC and the VRN took part in signing these letters (Save the Mekong 2010, 2011, 2012a; 263 NGOs 2011). This illustrates a similarity in the letter-writing approaches of both coalitions. However, differences in the approaches of these networks were also observed while examining the case studies. Among four of the letters which were addressed to regional decision-makers during the study period, the Vietnamese coalition signed all the letters on behalf of its coalition, whereas the Cambodian coalition had different signatories to the letters each time. The first letter was signed by the coalition as a whole and the remainder were signed by some individual member NGOs, not by the full membership (Save the Mekong 2010, 2011, 2012a; 263 NGOs 2011).

The first letter was drafted by the members of the STM Coalition in October 2010, when the PNPCA process officially commenced (Save the Mekong 2010). The letter called for a halt to the PNPCA process and for cancellation of the Xayaburi dam (Save the Mekong 2010). It was addressed to the CEO of the MRC and copied to the members of the Joint Committee (JC) to the MRC, and to the MRC's donor governments. The letter included two main reasons for this request, both relating to the deficiencies in the 1995 Mekong Agreement and the PNPCA. First, it was claimed that the Xayaburi dam project documents submitted to the MRC by the government of Laos had not been released to the public, leading to a lack of transparency in the process (Save the Mekong 2010: 1; MRC 2011). Second, the letter pointed out that the Xayaburi PNPCA process had started without waiting for the final release of the strategic environmental assessment (SEA) of the mainstream hydropower dams on the Mekong River, which was commissioned by the MRC (Save the Mekong 2010: 2). Both arguments claim that there was a lack of clear process, and thus question the legitimacy of the way the PNPCA was conducted. This appears to be ultimately caused by the deficiency of the 1995 Mekong Agreement, once again illustrating the influence of this formal rule at the regional level.

The second letter was drafted by the STM Coalition in January 2011, prior to the MRC Council meeting. This time, the letter was addressed to the members of the MRC Council from each MRC member country. In a similar way to the previous letter to the CEO of the MRC, this letter requested a halt of the Xayaburi PNPCA process. This process was at the time already four months long, and was being conducted without official endorsement of the SEA report commissioned by the MRC, and without disclosure of relevant project documents to the public (Save the Mekong 2011). Both the first and second letters claimed that there were deficiencies in the PNPCA process. The third letter was signed by 263 NGOs from 51 countries in March 2011. It was addressed to the Prime Ministers of Thailand and Laos, and requested the cancellation of the Xayaburi dam (263 NGOs 2011). The letter was signed by the VRN and some of the member NGOs of the RCC.

The fourth letter was issued on 20 April 2012, requesting clarification on the prior consultation of the Xayaburi dam. This time, the same letter was sent individually to the CEO of the MRC, and to the council members of the National Mekong Committees of Thailand, Cambodia, and Vietnam. The letters were written while construction was ongoing at the Xayaburi dam site. While the CH. Karnchang company, the contractor for construction of the dam, declared that it had already signed the agreement on the construction of the Xayaburi dam (CH. Karnchang 2012; Vandenbrink 2012), representatives of the MRC had agreed to conduct further studies on the impact of mainstream dams (MRC 2012). For Cambodia, the letter was addressed to Lim Kean Hor, the Minister of Water Resources and Meteorology and the Chairperson of the Cambodian National Mekong Committee (Save the Mekong 2012b). As in the case of the second letter, the signing of this letter by the NGOs was prompted by perceived deficiencies in the PNPCA process. This was seen as causing ambiguity in the interpretation of when the PNPCA process could be considered complete.

Interviewees from both the RCC and VRN suggested that while the STM Coalition does not have a formal coordinator, International Rivers (IR) and the Towards Ecological Recovery and Regional Alliance (TERRA) function as informal coordinators to the network (V2 2012; C5 2012; V11 2012). Letters addressed to regional decision-makers were drafted by these coordinating NGOs and circulated to the members within the STM Coalition (C5 2012). The main task of the RCC and VRN members was to comment on the content of the letters and, in the case of Cambodia, to determine whether to sign the letter as the RCC coalition, or to sign as an individual NGO. According to one STM interviewee, the language used in the letter was the most difficult thing to agree upon among the members, as some members preferred a softer approach than others. Agreeing on a letter such as this could take approximately three to four weeks (R2 2012).

In addition to these letters initiated by the regional STM Coalition, the RCC wrote a letter of its own in July 2012. It was addressed to the Prime Ministers of Thailand and Laos in response to the actions of regional actors such as the Xayaburi Power Company, the government of Laos, and NGOs in Thailand. It was the first letter addressed to the regional-level decision-makers written by the RCC in its own right, rather than as part of the STM Coalition. The letter requested cancellation of Thailand's power purchase agreement from the Xayaburi dam, a stop to the construction of the Xayaburi dam, and that the 1995 Mekong Agreement be respected (NGO Forum on Cambodia 2012). It was written while Laos officially admitted the continuation of Xayaburi construction activities, despite there being no consensus among the four countries over the Xayaburi dam (International Rivers 2012). At the same time, a civil society group in Thailand had just filed a court case to the Thai administration court demanding cancellation of the Electricity Generating Authority of Thailand (EGAT)'s power purchase agreement with the Xayaburi dam company (Wipatayotin 2012). The RCC also posted its message on the opinion column of the *Bangkok Post*, targeting Thai audiences (Ath and Phalika 2012).

The RCC also worked with community members along the Mekong River to collect thumb print petitions requesting a stop to the Xayaburi dam. Thumb prints are used in these cases to endorse signatures. The petitions were sent to the Prime Ministers of Thailand and Laos (NGO Forum on Cambodia 2012). Thumb print petitions were collected and sent to regional decision-makers on two occasions: the first in December 2011, the month when the MRC member governments met at the MRC council meeting; the second in June 2012, during a peace walk organized in Kampong Cham province, one of the Cambodian provinces along the Mekong River (NGO Forum on Cambodia 2012).

There was a difference in how network members viewed activities targeting regional audiences. During the author's field interviews, only one respondent mentioned this type of activity in Vietnam, whereas several Cambodian respondents indicated participation in regional activities such as taking part in signing regional letters as one of the main activities of the RCC (V2 2012; C3 2012; C7 2012; C13 2012). These comments reflect differences in the relative importance of targeting regional decision-makers by the RCC as compared with the VRN.

In June 2012, an international coalition of civil society groups, coordinated by the Siemenpuu Foundation, a Finnish NGO, filed a complaint concerning Pöyry and sent it to the Organisation for Economic Co-operation and Development (OECD) national contact point (OECD Watch 2013; Siemenpuu Foundation Mekong Group 2012). The OECD *Guidelines for Multinational Enterprises* (OECD 2011) allow citizens to file a complaint on the operations of multinational corporations based in OECD nations through designated national contact points in each OECD country (Oshionebo 2013). Using this scheme, the civil society coalition coordinated by the Siemenpuu Foundation complained that the advice provided by Pöyry to the Lao government undermined the cooperative regional process of the Mekong River (OECD Watch 2013). Both the VRN and the RCC took part in this case (OECD Watch 2013). However, during the author's fieldwork, only one interviewee from the VRN mentioned this activity (V10 2012), which raises the question of the activity's importance for the VRN and RCC.

Comparative analysis

The analysis of the VRN's and RCC's strategies illustrates similarities and differences in the approaches taken by both networks. The analysis of the interview data, based on the grounded theory approach discussed in Chapter 3, identified themes that were repeated by the interviewees. Two key themes emerged from these interviews concerning the strategies used in targeting regional decision-makers. These themes include: 1) the relationships between the national and regional coalitions; and 2) network culture. The following sections discuss these themes in detail and identify how rules and norms influenced the NGO coalitions' strategies.

Relationship between the national coalitions and the regional coalition

There are similarities and differences in the RCC's and VRN's approach in targeting regional-level decision-makers. The analysis of previous studies and the interview data indicate that historical engagements and interactions between national coalitions and the regional coalitions contribute to shaping their strategies. The following sections first look at the RCC in Cambodia, followed by an analysis of the VRN in Vietnam.

The Cambodian coalition, in its early days when it was called the Se San working group, focused on supporting Cambodian communities impacted by the Yali dam built and operated in the Vietnamese part of the Se San River, a tributary of the Mekong River (NGO Forum on Cambodia undated). As discussed in Chapter 4, the Yali dam is a Vietnamese dam built close to the border of Cambodia. This dam impacted negatively on the livelihoods and well-being of the downstream Cambodian communities, creating floods and causing contamination of river water (Wyatt and Baird 2007). The MRC was not effective in resolving the issues that the local communities in Cambodia were facing (Thim 2010). The Se San working group was established through activities focusing on the transboundary impacts of the Yali dam. The Se San working group was established through a network of local communities, local NGOs, and international NGOs aimed at supporting the affected communities; this was the predecessor of the RCC (C11 2012). Some of the key international supporters during this initial phase included Oxfam America, TERRA, IR, the Australian Mekong Resource Centre (AMRC), Probe International, and Mekong Watch (C11 2012). According to interviewees who were engaged in the network from its early stages, the formulation of the network was driven not only by the needs of the affected communities, but also through the interests of external partners (C11 2012; C28 2012). A number of partners collaborated in these activities using their specific strengths and interests: for example, Oxfam America provided technical and funding support for the network; TERRA wrote a letter to the MRC posing questions on the Yali dam; and IR together with Probe International contacted the World Bank on the issue of the transmission line from the Yali dam (Thim 2010: 164).

This style of working reflects the fact that the RCC worked in the style of a TAN from an early stage. Each partner used their own contacts and resources to pressure different actors who could potentially influence the decisions related to the Yali dam and further plans for dams on the Se San River. Support from international actors was critical in the establishment and continuation of the network. In addition, the transboundary nature of the Se San River and the Yali dam encouraged the RCC to look to international support at an initial stage, and the Cambodian NGOs benefited from engaging with international actors at an early stage. This international outlook has continued in the RCC;

the VRN, on the other hand, had limited international focus until the case of the Xayaburi dam (Vietnam Rivers Network 2013; V2 2012).

Funding for the Se San network originally came from Oxfam America, and when Oxfam made a decision to stop further funding support due to changes in priorities by Oxfam America, the network faced difficulties in continuing its work (C28 2012; C11 2012). This demonstrates the resource dependency of the Cambodian network. While the RCC network's decision-making process is currently based on member consensus (this point is discussed further in the next section), interviewees indicated the continuing influence of regional partners in the RCC's activities, particularly through IR, which often provides suggestions for the RCC's work associated with hydropower dams (C9 2012; C6 2012; C13 2012). An example of this influence was illustrated through a comment of one interviewee, who stated that the idea of writing the RCC's own letter in 2012 was suggested by a member of staff from the IR who used to be an advisor to the RCC (C6 2012). Several other RCC members commented that some of their activities are funded by IR and that they are decided through discussions between IR and the member NGOs (C9 2012; C22 2012). Referring back to the discussion of political responsibility within TANs suggested by Jordan and van Tuijl (2000) in Chapter 2, it would seem that the relationship between the RCC and the STM Coalition is close to what Jordan and van Tuijl describe as a cooperative campaign relationship (Jordan and van Tuijl 2000).

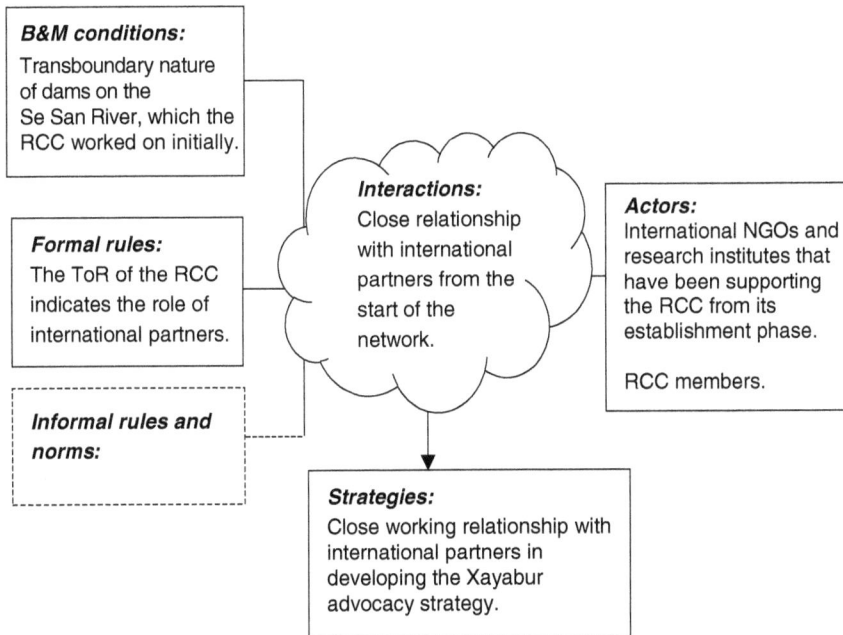

Figure 8.1 Influence of formal rules and biophysical and material conditions on the RCC strategy for close collaboration with international partners

In summary, two main factors were important for the RCC's strategies. One was the interaction among actors, i.e. close working relationships and resource dependency with the RCC's international partners. Second, the physical and material condition of the hydropower dam issues on the Se San River, with its transboundary nature, created a situation where it was essential for the RCC to collaborate closely with international partners from its early days. This close relationship with international partners is reflected in the RCC's ToR, which is the formal rule established by the members of the coalition. The older version of the ToR, which was in effect during the fieldwork for this book, identifies these international partners as supporting organizations that 'wish to support the efforts and work of the RCC but are unable to adhere to the roles and responsibilities of the core member organizations' (Rivers Coalition in Cambodia undated). The new ToR of the RCC changes the status of international partners as it identifies 14 international organizations as 'members' of the RCC (Rivers Coalition in Cambodia 2012). These factors affected the RCC's relationship with the STM Coalition members, resulting in the RCC's strategy of close collaboration with international partners. Figure 8.1 is an application of this book's analytical framework to this analysis. Where the analysis did not find influence of certain factors over the advocacy strategies, this is illustrated with dotted lines.

The VRN is also part of the STM Coalition, the same TAN in which the RCC takes part. The VRN has close relationships with IR, which provides the informal role of coordinating the STM Coalition. The IR commissioned the establishment of the VRN and continues to provide both technical and funding support to it (Vietnam Rivers Network 2013; R2 2011; V2 2012). However, comments from some of the network members below illustrate differences in approach between the VRN and its regional partners (IR and the STM Coalition, in this case).

> You see, the advocacy approach we selected is very appropriate in Vietnam. If we had worked like IR then we would have been killed by now.
>
> (V16 2011)

> IR and other regional groups know that they cannot come here and do advocacy because they do not have local context and knowledge, and because it is not a good model. It does not build longer term movement here.
>
> (V11 2012)

> We talk to the STM Coalition frankly that each country works differently, because we have different socio-political situations. So the approach in Thailand cannot work in Vietnam. At that time, many Thai NGOs said that you are not an NGO. They said that. But I said to them that we have an objective. Our belief is that how to deliver objective. So we have to investigate the best way to achieve our objective. So it is our approach, our

belief. I think up to now, they have to agree with us. So we join regional approach. But sometimes we did not join because it is not appropriate for us. Sometimes we have to stay outside. But we participate in developing common messages.

(V2 2011)

These comments reflect that while the VRN takes part in the STM Coalition, the VRN needs to be careful to take action in an appropriate manner within Vietnamese society. This highlights the importance of taking an embedded approach and the way that informal rules and norms affected this VRN member's attitude when working with the STM Coalition.

Another difference between the VRN and the RCC is that the Xayaburi dam was the first international river on which the VRN conducted advocacy work (Vietnam Rivers Network 2013). All the previous VRN work focused on Vietnamese domestic rivers (V11 2012). These activities involved monitoring the public consultation processes for the Trung Song hydropower project and the Song Bung 4 hydropower project, both located in the central highlands in Vietnam. They also raised concerns on the Dong Nai hydropower dams located in southern Vietnam (Vietnam Rivers Network 2013: 10). The network also provided comments on the new law on water resources, and worked with the National Assembly to embed the integrated water resources concept into this law (Vietnam Rivers Network 2013: 10). The VRN's work with domestic dams was driven primarily by the VRN members themselves, and none of the interviewees mentioned the involvement of international partners other than through funding.

The VRN's independent working style is stated in its strategy for 2008–10 as:

Different from majority of other networks established by donor-funded projects, VRN was initiated by a group of passionate environmental activists in Vietnam and expanded following the rapid and unsustainable development of hydropower in the country. From its very beginnings, the network was self-initiated and self-led and is not reliant on one certain donor.

(Vietnam Rivers Network 2009b)

This approach was reflected in the work of the VRN when it was engaged with the advocacy work on the Xayaburi dam, which was driven by the VRN's Mekong task force. This task force consisted of a small number of Vietnamese members, including scientists based in the Mekong Delta, and other members based in Hanoi whose roles were primarily to coordinate the task force's work and to liaise with various actors including decision-makers and partners at national and international levels (V1 2011; V2 2012; V5 2012; V10 2012). This comparatively independent working style of the VRN has potentially shaped the differences observed in the work on the Xayaburi dam of the VRN and its regional partners.

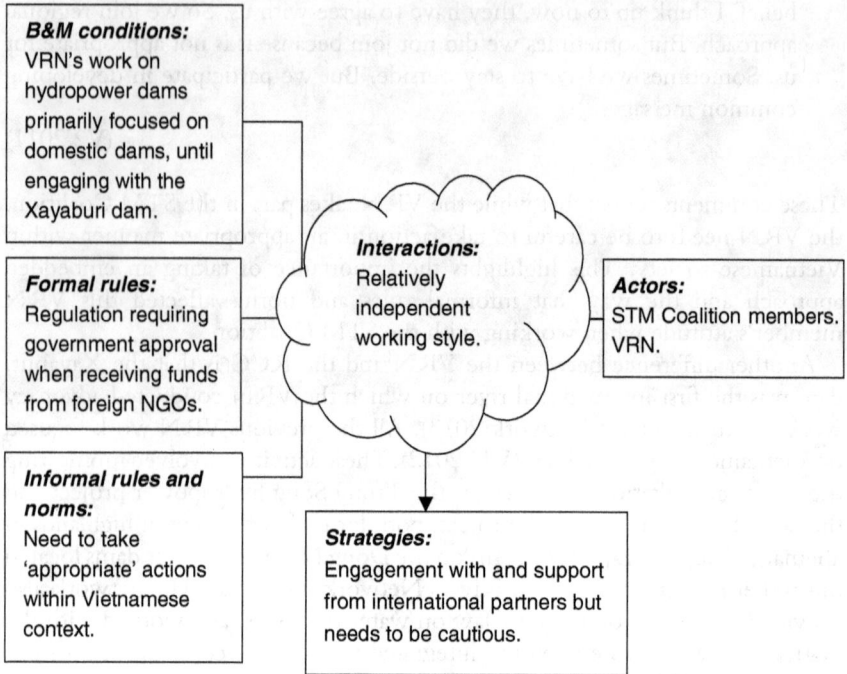

Figure 8.2 Influence of norms, actors, biophysical and material conditions on the VRN's strategy in engaging with regional partners

The way different factors affected the VRN's strategy is illustrated in Figure 8.2. While the VRN has an ongoing working relationship with other STM Coalition members from its establishment phase (particularly with the IR), the informal rules and norms that require Vietnamese NGOs to behave appropriately within their cultural and political context make VRN members take a cautious approach when engaging with regional actors. As discussed in Chapter 6, the Vietnamese government's regulations require that aid money provided by foreign NGOs must be approved by government agencies (*Regulation on the Use of Aid from INGO* 2009). While none of the interviewees indicated any specific influence of this regulation on the VRN's strategy targeting regional decision-makers, such a formal rule presumably contributes to the cautious approach adopted by the VRN in working with international partners.

In addition, the hydropower dams on which the VRN worked in its establishment phase are primarily domestic dams. Although this point was not mentioned in particular by any of the interviewees, this influence of physical and material conditions of the advocacy subject presumably did not provide a strong motivation for the VRN to collaborate with international partners in the past. These factors contributed to the VRN's approach when working with regional partners.

Network culture

The way the letters were signed can be seen to reflect the influence of internal rules and norms within the NGO coalitions. The first letter in October 2010 was signed by the RCC, whereas the letters in January 2011 and April 2012 were signed by only some of the RCC members (Save the Mekong 2010, 2011, 2012b). According to a key member of the RCC, this difference was caused by the fact that it was not possible to gain consensus among some of the RCC member organizations within the limited period of time necessary for the signing of the letter (C5 2012). Another member indicated that there was hesitancy among some member organizations, especially in Cambodia, towards signing petitions which may be considered as 'shaming' governments. This is another possible reason for the limited signing of the letter (C28 2012).

As discussed in Chapter 6, decision-making within the RCC is determined by its ToR (Rivers Coalition in Cambodia undated). According to the RCC's ToR, its regular decisions are made by consensus. However, when consensus cannot be achieved, certain member organizations designated as core members can make a decision based on majority vote (Rivers Coalition in Cambodia undated). When the RCC attempted to take part in signing letters drafted by the STM Coalition, the RCC members at times could not reach agreement on whether or not to sign the letters as a coalition, leaving some of the most positive member organizations to sign the letters individually. The procedure for signing the letters illustrates the influence of the internal formal rules adopted by the members of the RCC. An interviewee indicated that this formal style of decision-making started when the network expanded from the 3S network to the RCC, expanding its geographic scope as well as its membership base (C28 2012). The same interviewee continued that, as it became a wider network, and with adoption of formal rules, the decision-making process became slower, hindering the swift action needed in conducting advocacy (C28 2012).

On the other hand, the VRN has less formality in its decision-making (V5 2012). All the VRN's activities associated with the Xayaburi dam were led by the Mekong task force, which held decision-making authority (V2 2012). The task force, being a rather small group of individuals, could obtain consensus among the group on actions without a cumbersome process. As discussed in Chapter 6, the VRN does not have a strict decision-making rule (Vietnam Rivers Network 2009a; V5 2012), and when using the VRN's name and logo, one of the members of the executive committee needs to be involved in the activity (Article 21 Vietnam Rivers Network 2009a). Since the VRN's Mekong task force included a member of the executive board, the group did not have to gain permission from the executive board each time they made a decision on Xayaburi advocacy work.

Another factor that affected the VRN's decision-making process is its working culture within the coalition. As one of the interviewees commented,

The success of this group is that firstly, we have very strong commitment, and common interest, vision, mission to protect the Mekong Delta. That one leads us to cooperate together. So we get very easy to agree with each other, on what we need to do and who will do what. I don't think we have debate or conflict on doing things.

(V2 2012).

This comment reflects the VRN's voluntary working culture, a result of being a coalition of members voluntarily engaged in the network. This voluntary culture based on individual commitment is contrasted with the situation of the RCC, where most members take part in the coalition's activity as part of their jobs.

From a practical point of view, the current decision-making process may also be associated with the nature and the size of the coalition memberships. The total number of the RCC membership is fewer than 30 organizations, which makes consensus-based decisions more feasible. In contrast, the VRN has approximately 300 members, which makes it difficult to have a consensus-based decision-making system by all members (Vietnam Rivers Network 2012).

As illustrated in Figure 8.3, the analysis indicates that formal rules established by each coalition, along with the nature of network membership, affected the working culture within the case study coalitions. This working culture affected the way each coalition was able to sign the letters drafted by the STM Coalition. This analysis illustrates that interaction of formal rules, actors, and informal rules and norms, created patterns of interactions that affected the way each coalition signed the letters targeted at regional decision-makers.

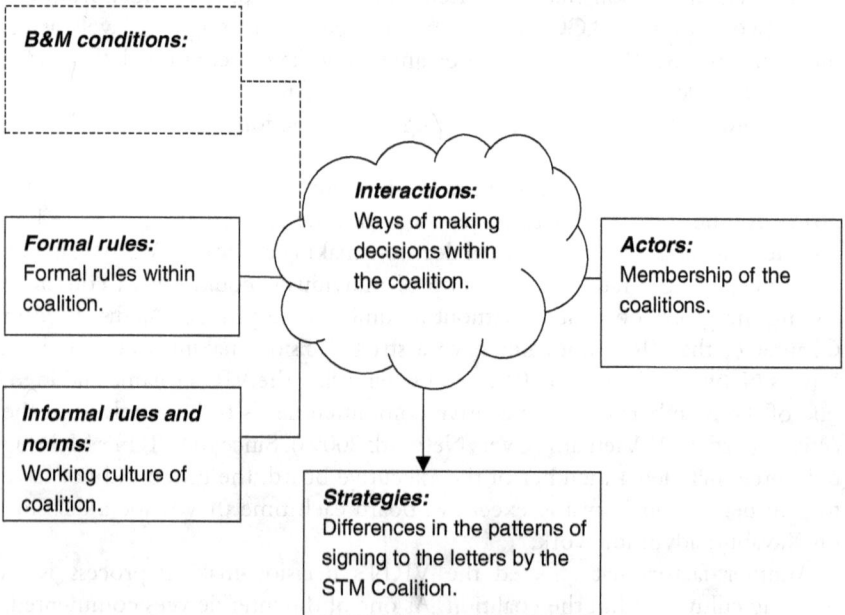

B&M conditions:

Interactions:
Ways of making decisions within the coalition.

Formal rules:
Formal rules within coalition.

Actors:
Membership of the coalitions.

Informal rules and norms:
Working culture of coalition.

Strategies:
Differences in the patterns of signing to the letters by the STM Coalition.

Figure 8.3 Influence of formal rules, informal rules and norms, and actors on strategy for signing letters targeting regional decision-makers

Conclusion

The analysis of the RCC's and VRN's strategies for targeting regional decision-makers provides a number of answers to the question of how rules and norms influence the regional advocacy strategies of NGO coalitions. There are three key findings from this chapter. The first is that historical relationships among NGO actors influence the dynamics of these actors' relations, which in turn affect the way strategies are shaped. This was illustrated through an analysis of the relationships between the national and regional coalitions. This relationship also reflects the nature of the TAN, where network members are resource-dependent on each other. In the case of Vietnam, the norm that requires NGO coalitions to act appropriately also has an important influence on this relationship.

Second, formal rules within the coalition and the informal working culture within the coalition, together with membership characteristics, all contribute to the ways decisions are made within the coalition. This decision-making mechanism, which is created as a result of the interaction of formal rules, informal rules and norms, and actors, affects the way strategies are undertaken by both coalitions.

Finally, ambiguity in formal rules such as the 1995 Mekong Agreement and the PNPCA created the situation where MRC member states understood the process of consultation through their own interpretation of requirements. This led to the NGO coalitions' strategy at the regional level focusing on a criticism of the MRC member states' way of implementing the PNPCA.

References

263 NGOs. 2011. *Global Call to Cancel the Xayaburi Dam on the Mekong River mainstream in Northern Lao PDR.* A letter addressed to Thongsing Thammavong, the Prime Minister of Lao PDR and Abhisit Vejjajiva, the Prime Minister of Thailand.

Ath, Chhith Sam, and Chea Phalika. 2012. A plea for Xayaburi. *Bangkok Post:* Opinion, 5 August. www.bangkokpost.com/opinion/opinion/306019/a-plea-for-xayaburi. Accessed 28 September 2012.

C3. 2012. Personal interview, 25 July 2012.

C5. 2012. Personal interview, 25 July 2012.

C6. 2012. email communication to the author, 26 July 2012.

C7. 2012. Personal interview, 27 July 2012.

C9. 2012. Personal interview, 28 July 2012.

C11. 2012. Personal interview, 28 July 2012.

C13. 2012. Personal interview, 31 July 2012.

C22. 2012. Personal interview, 7 August 2012.

C28. 2012. Personal interview, 10 August 2012.

CH. Karnchang 2012. *Notification of Contract Signing with Xayaburi Power Company Limited.* www.ch-karnchang.co.th/news_activities_detail_en.php?nid=474. Accessed 19 October 2012; no longer available online.

International Rivers. 2012. Testing the waters: Laos pushes xayaburi dam to critical point. www.internationalrivers.org/blogs/267/testing-the-waters-laos-pushes-xayaburi-dam-to-critical-point. Accessed 19 October 2012.

Jordan, Lisa, and Peter van Tuijl. 2000. Political responsibility in transnational NGO advocacy. *World Development* 28 (12): 2051–2065.

MRC. 2011. *Prior Consultation Project Review Report: Volume 2 Stakeholder Consultations related to the proposed Xayaburi dam project*. Mekong River Commission. www.mrcmekong. org/pnpca/2011-03-31-Report-on-Stakeholder-Cons-on-Xayaburi.pdf. Accessed 11 April 2011.

—— 2012. *Minutes of the Eighteenth Meeting of the MRC Council*, 8 December 2011, Siem Reap, Cambodia. www.mrcmekong.org/assets/Publications/governance/Minutes-of-the-18th-Council.pdf. Accessed 29 September 2012.

NGO Forum on Cambodia. 2012. *Subject: Rivers Coalition in Cambodia Joint Statement: Call Thailand to Cancel the Power Purchase Agreement and Laos to stop construction and for both countries to respect the 1995 Mekong Agreement*. A letter addressed to H.E. Yingluck Shinawatra, Prime Minister of Thailand and H.E. Thongsing Thammavong, Prime Minister of Lao PDR, 27 July.

—— undated. *Hydropower and Community Rights Project*. www.ngoforum.org.kh/eng/en_project_artticle.php?artticle=8. Accessed 13 October 2012; no longer available online. Updated version at www.ngoforum.org.kh/index.php/en/community-rights-on-hydropower-development-project. Accessed 7 June 2015.

OECD. 2011. *OECD Guidelines for Multinational Enterprises*. 2011 Edition. Paris: Organisation for Economic Co-operation and Development. www.oecd.org/daf/inv/mne/48004323.pdf. Accessed 2 May 2013.

OECD Watch. 2013. *Siemenpuu et al vs Pöyry Group*. http://oecdwatch.org/cases/Case_259. Accessed 4 October 2013.

Oshionebo, Evaristus. 2013. *Community Remedies under the OECD Guidelines for MNEs*. Presentation at the Community Rights and Expectations in Natural Resources Development seminar, University of Dundee, 24–25 April 2013.

R2. 2011. Personal interview, 16 November 2011.

—— 2012. Personal interview, 30 June 2012.

Regulation on management and use of foreign nongovernmental aid promulgated by Decree No.93/2009/ND-CP of 22 October 2009 of the Government 2009. Vietnam.

Rivers Coalition in Cambodia. 2012. Rivers Coalition in Cambodia (RCC) Terms of Reference (document obtained from an interviewee).

—— undated. Rivers Coalition in Cambodia (RCC) Terms of Reference (document obtained from an interviewee).

Save the Mekong. 2009. *About Save the Mekong Coalition*. www.savethemekong.org/issue_detail.php?sid=13. Accessed 19 October 2012.

—— 2010. *Subject: Call for Halt to the PNPCA Process and Cancellation of Xayaboury Dam*. A letter to Mr Jeremy Bird, CEO, Mekong River Commission, 13 October.

—— 2011. *Subject: Request for MRC Council to halt the current PNPCA process on the Xayaburi Dam, to endorse the MRC's SEA report's findings, and commit to evaluate all options for meeting the Mekong region's water and energy needs through a credible and objective public process*. A letter addressed to H.E. Mr Lim Kean Hor, Minister of Water Resources and Meteorology *et al.*, 25 January.

—— 2012a. *Subject: Request for Clarifications on the Prior Consultation for the Xayaburi Dam*. A letter addressed to Mr Hans Guttman, CEO, the Mekong River Commission, 20 April.

—— 2012b. *Subject: Request for Clarifications on the Prior Consultation for the Xayaburi Dam*. A letter addressed to Mr Lim Kean Hor, Minister of Water Resources and Meteorology, Cambodia, 20 April.

Siemenpuu Foundation Mekong Group. 2012. *Specific Instance to OECD National Contact Point in Finland: the role of Pöyry Group services in the process of the Xayaburi hydropower project in Lao PDR*. http://oecdwatch.org/cases/Case_259/1040/at_download/file. Accessed 5 May 2014.

Thim, Ly. 2010. Dynamics of planning process in the Lower Mekong Basin: a management analysis for the Se San Sub-Basin. PhD thesis, Philosophischen Fakultät, Rheinischen Friedrich-Wilhelms-Universität, Bonn.

V1. 2011. Personal interview, 10 November 2011.

V2. 2011. Personal interview, 23 November 2011.

—— 2012. Personal interview, 10 July 2012.

V5. 2012. Personal interview, 11 July 2012.

V10. 2012. Personal interview, 14 July 2012.

V11. 2012. Personal interview, 14 July 2012.

V16. 2011. Personal interview, 23 November 2011.

Vandenbrink, Rachel. 2012. Cambodia Lodges Dam Protest with Laos. *Radio Free Asia*, 1 May. www.rfa.org/english/news/laos/xayaburi-05012012190456.html. Accessed 3 May 2012.

Vietnam Rivers Network. 2009a. *Regulations of Vietnam Rivers Network*. http://vrn.org.vn/en/h/d/2012/04/254/How_can_I_register_for_VRN_membership/index.html. Accessed 18 February 2013.

—— 2009b. Vietnam Rivers Network's Strategy 2008–2020. http://vrn.org.vn/en/h/d/2012/04/245/VRN%27s_strategy_2008-2020/index.html. Accessed 7 July 2015.

—— 2012. *About Us / Introduction*. http://vrn.org.vn/en/h/d/2012/04/244/Introduction/index.html. Accessed 7 November 2012.

—— 2013. *Báo Cáo Tổng Kết Hoạt Động* [Activity Report] 2006–2012. http://vrn.org.vn/media/files/VRN%20annual%20report_Final.pdf. Accessed 12 February 2013.

Wipatayotin, Apinya. 2012. Thais ask court to stop Xayaburi dam project. *Bangkok Post*, 8 August. www.bangkokpost.com/news/local/306554/thais-ask-court-to-stop-xayaburi-dam-project. Accessed 29 September 2012.

Wyatt, Andrew B., and Ian G. Baird. 2007. Transboundary impact assessment in the Sesan River Basin: the case of the Yali Falls dam. *International Journal of Water Resources Development* 23 (3): 427–442.

9 Strategies targeting national decision-makers

Introduction

As discussed in Chapter 6, the decision-making process under the 1995 Mekong Agreement over the use of the Mekong River is based on consensus among the four member states (Articles 20, 27 *Mekong Agreement* 1995). Therefore the national governments' positions over the Xayaburi dam are the key determining factor affecting the fate of the regional decision on the dam. Consequently, influencing national-level decision-makers became one of the most important advocacy strategies for both NGO coalitions.

During the Xayaburi Procedures for Notification, Prior Consultation and Agreement (PNPCA) process, both the Rivers Coalition in Cambodia (RCC) and the Vietnam Rivers Network (VRN) conducted activities aimed at influencing their respective governments' decision-making processes. The main strategies included conducting workshops targeting national decision-makers; providing direct inputs to government decision-makers through formal and informal meetings; and using science as a basis for the dialogue with decision-makers. This chapter analyses and compares the strategies adopted by the RCC and VRN, and attempts to identify how rules and norms may have shaped these strategies. First, the chapter compares how and why the RCC and VRN adopted the strategies; the following section examines key factors shaping these strategies, aiming to identify how rules and norms influenced the advocacy strategies of NGO coalitions.

Strategies adopted by the VRN

From the start of the PNPCA process, targeting national decision-makers was the VRN's main strategy, and it used several tactics in its attempts to influence the national decision-making process. First, the VRN organized various workshops aimed at relatively focused audiences, which included national decision-makers, scientists, and the media. Second, VRN members used both formal and informal pathways to directly approach their target audiences. Third, science played an important role in their advocacy work. This section discusses the VRN's activities in detail, leading to a further analysis of the relationship between the VRN's activities and rules and norms in the following section.

Workshops

During the Xayaburi hydropower dam PNPCA process, the VRN organized several workshops targeting decision-makers, scientists, and media. Many of these workshops were co-organized with other partners, particularly with government institutions including the Vietnam Union of Science and Technology Associations (VUSTA) and the Vietnam National Mekong Committee (VNMC). The workshops were targeted at certain types of audience, primarily national-level decision-makers, scientists, and media.

The VRN started its work targeting national decision-makers at an early stage of the PNPCA process. On 7 November 2010, the month after the PNPCA process started, the VRN and Pan Nature, a Vietnamese NGO, co-hosted a public dialogue on the development of Mekong mainstream dams and their implications for Vietnam (Dien 2010; V3 2012; V16 2012). This dialogue was primarily targeted at National Assembly members and VRN members. During this dialogue, many of the National Assembly members proposed a public hearing on the Mekong mainstream dam (Dien 2010). Another workshop was organized in October 2011 by Pan Nature, targeted at National Assembly members in Hanoi, on issues including the Xayaburi dam, where the VRN members collaborated as a resource (V20 2011).

As discussed in Chapter 5, the National Assembly is the highest decision-making body representing citizens' interests in the Vietnamese political system (Article 6 *Constitution of Vietnam* 2001). The collaboration with the members of the National Assembly was possible partly due to the ongoing working relationship between Pan Nature and the National Assembly on natural resources-related issues since 2007 (V3 2012).

The VRN conducted workshops at certain key junctures of the PNPCA process. On 15 March 2011, a month after the official consultations of the PNPCA were conducted by the VNMC, key members of the VRN and the Vietnam Water Partnership organized an open dialogue on the gains and losses of the Xayaburi dam for Vietnam (V16 2012; VUSTA *et al.* 2011; Quới 2011). It was held in Hanoi, and 60 representatives from the government, civil society, and scientists participated in the dialogue (Trandem 2011). The workshop was opened by a representative of VUSTA, and technical inputs associated with the workshop subject were provided by a member of the VNMC, as well as by individuals from within the network of organizers (VUSTA *et al.* 2011; Quới 2011). The participants recommended ten years' deferral of the Mekong mainstream dams until academic research was conducted allowing decisions to take into consideration all the risks (Trandem 2011). This recommendation follows the recommendations of the strategic environmental assessment (SEA) commissioned by the Mekong River Commission (MRC) (ICEM 2010b). The Vietnamese government's official position over the dam, which was communicated to the MRC as a response of the PNPCA, also follows this recommendation of ten years' deferral (Socialist Republic of Viet Nam 2011).

On 23 November 2011, a few weeks before the MRC Council was due to meet in December 2011 when the regional decision was to be made on the Xayaburi dam, the VRN collaborated with the VUSTA in organizing a workshop on the Mekong's hydropower dams. The workshop was organized in Ho Chi Minh City, and was targeted primarily at scientists, government staff, media, and any other interested individuals (V7 2012). One focus of discussion at the workshop was the Pöyry report, which was commissioned by the government of Laos to review Laos' compliance with the Mekong Agreement, the MRC's design guidelines on sustainable hydropower, and concerns raised by other riparian countries over the Xayaburi dam (Government of Lao PDR and Pöyry 2011). The workshop's focus on the Pöyry report was in line with the concern of the Vietnamese government, which had just conducted a bilateral meeting with the government of Laos to discuss the report before this workshop took place (V16 2011).

The day before the workshop, the VRN organized a training workshop on coastal geomorphology and sediment transit in collaboration with the World Wide Fund for Nature (WWF) (Vietnam Rivers Network 2011b). Approximately 40 people participated in this workshop, including Vietnamese scientists and members of the VRN. The VRN members recognized the importance of increasing the number of 'allied' scientists. As one interviewee commented:

> There are scientists who do not understand the Delta and who support an engineering approach. They also provide inputs and the VUSTA consolidates.
>
> (V16 2012)

Another workshop was organized in Ho Chi Minh City in collaboration with the VUSTA on 14 August 2012, titled 'The Mekong and hydropower dams'. The workshop was attended by National Assembly members, members of mass organizations, representatives from 13 Mekong Delta provinces, journalists, academics, and international and national NGOs (V10 2012). The workshop took place amidst the situation in which the Lao government officially admitted that the construction of the Xayaburi dam was under way.

In summary, most of the VRN's national-level workshops were targeted at three main audiences: national-level decision-makers, scientists, and the media. Most of the workshops were conducted in collaboration with existing and new partners, including the VUSTA, Pan Nature, and the WWF. All the workshops had clear titles and themes discussing the Xayaburi dam or the dams on the Mekong River (Dien 2010; V3 2012; V16 2012; VUSTA *et al.* 2011), and no particular sensitivity was observed in raising the issue of hydropower dams on the Mekong River's mainstream.

Direct inputs

Another key strategy which the VRN adopted was to directly contact national decision-makers. The VRN used both formal and informal pathways in directly reaching these national-level decision-makers. This section provides an overview of how the VRN provided direct input to the decision-makers.

The main official pathway provided through the PNPCA process was the public consultation organized by each National Mekong Committee of the MRC's member countries. In Vietnam, the first PNPCA consultation was held in Can Tho city located within the Mekong Delta, inviting members of the research institutes based in Ho Chi Minh City, local governments' staff within the Mekong Delta, and members from Can Tho University (MRC 2011). The VRN Mekong task force members were not invited to this consultation (V2 2012; R2 2011; V25 2012) and the Hanoi based VRN members were informed about it only one day prior to the meeting, through their contacts in southern Vietnam (V2 2012). The VRN members requested that they be allowed to participate; however, they were not allowed to do so as the consultation was targeted at stakeholders within the Delta (V2 2012). During the consultation in Can Tho, VRN members communicated with some of the participants and managed to identify a few individuals willing to disseminate information regarding the Xayaburi dam (V2 2012). Some of the Hanoi-based VRN members were invited to the consultation workshop held at Halong Bay, organized primarily for Hanoi-based stakeholders (MRC 2011). During the consultation in Halong Bay, VRN participants brought information sheets on the Xayaburi and disseminated them to the participants (V2 2012). One of the VRN members, however, commented that it was not an activity that had any important influence (V2 2012).

Rather than using the official consultation meetings, the VRN's main strategy was to conduct informal meetings with key individuals who could convey their message to decision-makers (V2 2012; V16 2011). The VRN members attempted to identify a number of entry points to communicate their messages to decision-makers. One of the interviewees indicated that the VRN developed a matrix to identify allies and opponents to the VRN's opinion about the Xayaburi dam, a matrix that changed over time (V2 2012). As an example, the same interviewee indicated that in early 2010, the VNMC was not aligned with the VRN's position (V2 2012). However, by 2011 the VRN had gained sufficient trust with the VNMC, to the extent that the VNMC invited one of the VRN scientists to join the government delegation for a bilateral meeting with the government of Laos (V16 2011; V2 2011).

The VRN member indicated that informal meetings were one of the key activities they conducted as part of their Xayaburi advocacy work:

> Informal meetings are important in the context of Vietnam. We really have to push up the issue at national level. Because it is very difficult to

arrange meetings with senior ones, we have to raise the issue and see the one who can deliver the message to the target organization.

(V26 2012)

This comment reflects some Vietnamese scholars' claim that, in the Vietnamese political context, decisions are mostly made behind closed doors and public deliberations are rare (Cima 1987; Kerkvliet 2001) (discussed in Chapter 5). In this context, having an informal pathway to decision-makers becomes important.

Another interviewee also indicated that in 'Asian culture' people prefer to talk to each other informally before things become too formal (V2 2012). This reflects the regional political culture called the 'ASEAN way', which values personal relationships and seeks informal consultations and dialogues as a way of reaching consensus (Katsumata 2003: 104; Hirsch *et al.* 2006: 76).

Another important entry point for direct input was through the VUSTA, which is mandated to provide scientific advice to the Party and the State on policies related to national development (V7 2012; Article 6 *VUSTA Charter* 2012). With this mandate, the VUSTA was tasked by the government to provide scientific opinion on the Xayaburi hydropower dam to the Prime Minister and to other government offices during the Xayaburi PNPCA discussion process (V7 2012; V2 2012). VUSTA collected the opinions of scientists through workshops and direct inputs, and submitted its own report evaluating the Xayaburi dam's impact to the Vietnamese government (V7 2012). The VRN was able to use this channel effectively through providing direct input to VUSTA, as well as jointly organizing workshops (discussed above). This helped shape VUSTA's evaluation of the Xayaburi dam (V2 2012).

It appears that the VRN managed to gain the government's trust over the issue, and soon its members began to be invited to contribute to government-led initiatives. In November 2011, one of the scientists from the Mekong Delta who was also an active member of the VRN Mekong task force was invited to join the Vietnamese government's delegation to the bilateral meeting with the government of Laos to discuss the Pöyry report (V2 2011; V16 2011; V24 2013). It was the first time the VRN had been invited to the official international delegation. According to an interviewee, the invitation came from the government, which recognized the knowledge and expertise that existed within the VRN (V16 2012; V2 2011). This invitation was a significant achievement for the VRN, and one of the interviewees commented that:

Now our work is recognized by the VNMC. They share information with us, they invite us to give comments on issues, on the Pöyry report. It is very exceptional, because normally they share (information) within (the) government agency.

(V2 2011)

This relatively close relationship with the state organizations opened doors for the VRN to collaborate with other state organizations. The VRN was invited by the Ministry of Home Affairs and Security to brief its staff over the issue of the Xayaburi dam. This was a Ministry that does not have direct responsibilities associated with the Mekong River nor the Xayaburi hydropower dam, and thus was not the main target audience for advocacy strategies of the VRN. However, the Ministry was keen to learn about international water issues as they relate to water security (V2 2012).

The VRN also made direct inputs to decision-makers through official letters. On 15 April 2011, a few days before the MRC Joint Committee was scheduled to make an initial decision over the Xayaburi hydropower dam, the members of the VRN Mekong task force sent a letter of petition addressed to the key decision-makers within the government, including the Prime Minister, the Government Office, the Ministry of Natural Resources and Environment, the Ministry of Industry and Commerce, the Ministry of Agriculture and Rural Development, the VNMC, and VUSTA (WARECOD 2011). The letter provided an analysis of the impacts from the Xayaburi dam on the Mekong Delta, as well as its implications for neighbouring Laos and China (WARECOD 2011). It also provided a recommendation to the Vietnamese government, including the need for further impact studies from the 12 mainstream hydropower dams on the Mekong, before any projects could proceed (WARECOD 2011). The letter also recommends that the Vietnamese government supports Laos in exploring opportunities for developing dams on the tributaries of the Mekong River instead of the mainstream (WARECOD 2011). In May 2012, scientists from the VRN were reported to have made a direct request to the Prime Minister and the VNMC to stop the construction of the Xayaburi dam on the Mekong River's mainstream (*Thanh Nien News* 2012). This request came after the CH. Karnchang company, the contractor of the construction of the Xayaburi dam, officially admitted that the construction agreement was signed for the Xayaburi dam and that work had already begun (Dien 2012).

In summary, the VRN used a variety of methods to provide direct inputs to national decision-makers. The VRN did not place particular emphasis on the official PNPCA consultation; rather its strategy utilized informal meetings with various actors who could convey the message to decision-makers (V2 2012). These informal meetings relied on both personal contacts of the VRN members, and institutional linkages with VUSTA, whose formal role was in providing scientific advice to the government on its policy implications (V7 2012; Article 6 *VUSTA Charter* 2012). Informal contacts which VRN members brought to the organization's activities were valuable in providing informal channels to decision-makers. In an attempt to communicate with the decision-makers, VRN members tried to make arguments based on scientific evidence. The fact that some of the VRN Mekong task force members were scientists from the Mekong Delta supported this strategy. The next section discusses the VRN's science strategy in detail.

Use of science

According to a number of interviewees, the use of science was one of the key strategies adopted by the VRN (V16 2012; V2 2012; V17 2012). The group of scientists from the Mekong Delta, who were also members of the VRN's Mekong task force, was one of the main providers of scientific knowledge distributed through the network. These scientists from the Mekong Delta provided substantive inputs used in explanations of the potential impacts of the Xayaburi dam on Vietnam. These were presented during workshops with National Assembly members, other government officials, scientists, and community members (Tuan 2010; Thien 2012). The VRN also provided critical analysis of the Pöyry report, pointing out the weaknesses in its approach from a scientific point of view; these included aspects of water quality, sediment flow, and environmental impact assessment monitoring (Vietnam Rivers Network 2011a). The VRN members also translated the full SEA report into Vietnamese. The MRC officially translated only the summary of the SEA report but not the full text (MRC undated; V16 2012); this summary lacked detailed information about the Mekong Delta (V16 2012; ICEM 2010a). In addition, as discussed earlier, the VRN conducted a technical workshop to train scientists on hydropower dams' impacts on the Mekong Delta's geomorphology and sediments. This was organized in collaboration with WWF (Vietnam Rivers Network 2011b).

This strategy of using science was possible partly due to the inclusion of scientists from the Mekong Delta in the VRN Mekong task force. In addition, VUSTA created a pathway for the VRN to provide scientific information indirectly to the decision-makers. The advocacy strategy of the VRN at the national level can be summarized as using science, human networks, and workshops as key methods for achieving its advocacy targeted at national decision-makers. The next section discusses how the Cambodian network approached the same target audience in its country, and compares it with the strategy of the Vietnamese network.

Strategies adopted by the RCC

Compared with the VRN, which had its main focus on advocacy targeting national-level decision-makers, the RCC's activities targeting national-level audiences were rather limited. Two main strategies were adopted by the RCC: public forums and the use of formal meetings. This section describes the RCC's activities at this level in detail as a basis for analysing the relationship between rules, norms, and the RCC's strategies later in the chapter.

Public forums

During the period of study of this book, the RCC organized a public forum which included national decision-makers as its target audience. This forum was the National Conference on Climate Change, Agriculture and Energy, which was held on 1–2 December 2011. The conference was held in Phnom Penh and brought together 400 participants from the government, donors, communities, NGOs, and academics (NGO Forum on Cambodia 2011a). While the conference theme focused on climate change, the intention of the RCC in organizing this conference was to raise awareness of Mekong River hydropower dam issues (C5 2012b) prior to the MRC Council meeting in December 2011. The session on hydropower included presentations from the Cambodian National Mekong Committee (CNMC) representing the Cambodian government; the International Centre for Environmental Management (ICEM), the consultancy company that conducted the SEA of the Mekong River's mainstream dam; a scientist from the University of Aalto in Finland who studied the impacts of sediment; International Rivers; and community representatives who had been affected by hydropower development (NGO Forum on Cambodia 2011a). This conference was co-organized by the NGO Forum, the Council for Agriculture and Rural Development (CARD), and several other Cambodian NGOs (NGO Forum on Cambodia 2011a). CARD is under the Council of Ministers chaired by the Deputy Prime Minister, which coordinates and assists the Cambodian government on issues related to agriculture and rural development (CARD undated).

It was the first time that the hydropower programme of the NGO Forum collaborated with the Council of Minister's office (C5 2012b). Other programmes within the NGO Forum already had ongoing collaborations with the office of the Council of Ministers, and this existing connection allowed the hydropower programme to collaborate with high-level officials from the Cambodian government (C5 2012b).

In contrast to the VRN, which conducted several workshops targeting decision-makers such as the National Assembly members, the RCC's event targeted all types of participants, including national-level decision-makers, donors, communities, NGOs, and academics (NGO Forum on Cambodia 2011a). The event was held just before the MRC's council meeting, which was due to make a decision on the Xayaburi dam. The issue of the hydropower dam was not highlighted as a main topic of the event, but it was included as one of the themes (NGO Forum on Cambodia 2011a). According to one interviewee, this is related to the fact that the Cambodian NGO coalition felt it was too sensitive to discuss the hydropower dam with the Council of Ministers:

> It is not easy to talk about hydropower issue with Council of Ministers ....
> And this event is just awareness-raising, not a dialogue with the government.
> We just want to spread around the information about the hydropower in the
> Mekong, so when we try to put the agenda together, it's really soft. Not
> anything talking about the argument or any position or any perspective of
> community or civil society. So we try to make it soft, and try to communicate
> and explain more to the government, before we can organize such event.
>
> (C5 2012b)

Similar sensitivity in raising hydropower dam issues with local-level authorities was observed during the author's field interviews with RCC members based in the provinces (C12 2012; C13 2012), discussed more in detail in Chapter 10. This sensitivity at the national level is associated with the general informal pressure within Cambodian society not to pose questions against the development policy of the dominant political party, as discussed in Chapter 6.

Direct inputs

Another tactic adopted by the RCC in targeting national decision-makers was the use of formal processes. During the initial six months of the PNPCA process, public consultations were conducted by each National Mekong Committee. In Cambodia two public consultations were conducted: one consultation targeting local stakeholders who face potential impact from the Xayaburi dam; the other targeting national-level stakeholders. The RCC used these venues as a way to provide direct inputs to the government.

The first consultation was held on 10 February 2011 in Kratie, one of the provincial towns along the mainstream of the Mekong River (MRC 2011: 10). This first consultation was primarily for representatives from the six provinces along the Mekong River that are considered to be affected by the Xayaburi dam project. These include Stung Treng, Kratie, Kampong Cham, Kandal, Prey Veng, and Takeo provinces (MRC 2011: 10). Out of 68 participants at the Kratie consultation meeting, the majority of participants were provincial (26) and district (11) level government officers (MRC 2011: 25–27). Only four representatives of the communes, the lowest administrative level, attended the consultation. These were the heads of the communes from provinces along the Mekong River (MRC 2011: 25–27). Some RCC members based in these provinces were invited to the meeting. However, Phnom Penh-based RCC members were not initially invited. Having heard the news about the consultation from its provincial NGO members, the Phnom Penh-based RCC coordinator wrote a letter requesting CNMC to allow participation of other RCC members from Phnom Penh. This resulted in the CNMC allowing additional RCC members to join the consultation (C5 2011; C29 2012). In total, nine RCC members participated in the consultation, four from Phnom Penh-based NGOs (MRC 2011: 25–27). During this consultation, the members of the RCC distributed a Khmer-translated version of the fact sheet on the Xayaburi dam, which was originally published by International Rivers (C5 2011; International Rivers *et al.* 2011).

The second official consultation was held in Sihanoukville on 28 February 2011, this time targeting national-level stakeholders including government line agencies, NGOs, and research institutes (MRC 2011). Out of 43 participants at the consultation meeting, 25 were from the Cambodian national government agencies, and seven RCC members joined the meeting (MRC 2011: 28–29). The RCC requested a separate meeting with government officials prior to the consultation meeting (C5 2012a; C29 2011). This meeting was held during the consultation meeting in Sihanoukville, where a side meeting was held between the government representatives and RCC representatives. During this meeting, both sides could share their concerns over the Xayaburi dam (C5 2011; C29 2012). The RCC representatives at the meeting also expressed their wish to collaborate with the government on the overall dam issue (C5 2012a), resulting in a collaborative atmosphere and relationships between the CNMC and the RCC. Following this bilateral meeting, the RCC sent a thank-you letter to the CNMC emphasizing key important points from the meeting (NGO Forum on Cambodia 2011b). One government interviewee commented that this was the first time the government had received this type of appreciation from NGOs (C29 2011).

In contrast to the situation in Vietnam, the Cambodian NGO coalition did not use science as a strategy, nor did it conduct its own research related to the Xayaburi dam (C5 2012b). When necessary, the coalition used existing information from the MRC-commissioned SEA report. When the RCC needed technical inputs to the national-level workshops, external guest speakers were invited to provide the inputs.

Comparative analysis

The previous sections note the various strategies targeting national decision-makers that were adopted by the coalitions. For the VRN, targeting national decision-makers was the key focus of its advocacy strategy, and three key approaches were adopted: conducting workshops with targeted audiences; using formal and informal channels in order to contact decision-makers directly; and using science communications with decision-makers. In contrast, the RCC had somewhat limited activities targeting national decision-makers. The RCC approach was rather formal and used public forums, as well as using the PNPCA official consultation process organized by the government (CNMC). Why are there such differences in approach, and how do rules and norms play a role in characterizing each coalition's strategies? The following sections discuss key factors that shaped the two NGO coalitions' strategies, and analyse how different rules and norms influenced the NGO coalitions in shaping their strategies.

State–NGO relationships

A key theme that arises through comparing the two cases is the relationship between the state and NGOs. In Vietnam, the state–NGO relationship worked in favour of the VRN's Xayaburi strategy. This relationship supported the VRN in its strategy of: 1) the communication of science to decision-makers; 2) direct inputs to decision-makers; and 3) conducting targeted workshops. One of the most important relationships was with VUSTA. VUSTA created an official passage for the VRN to bring its message forward to high-level government officials, when VRN conducted targeted workshops, provided direct inputs, and used science as an entry point of discussion with the government (V2 2012; V7 2012). As discussed in Chapter 4, VUSTA is an umbrella organization for many Vietnamese NGOs (Norlund 2007: 9). The VRN is registered as one of the projects of the Centre for Water Resources Conservation and Development (WARECOD), a Vietnamese NGO registered under the auspices of VUSTA (Vietnam Rivers Network 2012; TERRA 2008: 32; WARECOD 2012).

Several formal rules supported the VRN's relationship with VUSTA, as illustrated in Figure 9.1. First, Decree 35-HDBT (1992) on the establishment of non-profit and science and technology organizations; Decree 81/2002/ND-CP on the implementation of the Science and Technology law; and Decree 88/2003/ND-CP (2003) on the organization, operation, and management of associations provide the legal basis for establishing Vietnamese NGOs as one of the science and technology organizations (*Decree No. 81/2002/ND-CP* 2002; *Decree No. 35/HTBT* 1992; *Decree No. 88/2003/ND-CP* 2003; Norlund *et al.* 2006: 72). Decree 30/2012/ND-CP on the organization and operation of social and charity funds includes an article that requires Vietnamese NGOs (VNGOs) to operate under recognized government agencies, including VUSTA (Article 4 *Decree No.30/2012/ND-CP* 2012; V10 2012). VNGOs registered with VUSTA are positioned as VUSTA's member associations (*VUSTA Regulation* 2006). The regulation of VUSTA indicates that these member associations have the right to participate in its activities (Article 9 *VUSTA Regulation* 2006). As defined through the Prime Minister's decision 22/2002/QD-TTg, issued in 2002, VUSTA is mandated to critically review and provide scientific opinions on government policy and projects (*Decision No. 22/2002/QD-TTg* 2002). This mandate allows VUSTA to fulfil a check-and-balance function for the Vietnamese government (Wells-Dang 2014; V7 2012; Huynh and Tuan 2007: 11). Through this mandate, VUSTA was tasked to provide scientific inputs associated with the Xayaburi dam to the Prime Minister's office. In conducting this task, VUSTA consulted with many of its member organizations, including the VRN, with which it organized joint workshops discussing the impacts of hydropower dams on the mainstream of the Mekong (V7 2012).

This relationship with VUSTA reflects the close state–civil society relationship in Vietnam. Many scholars discuss this close relationship and the 'fuzzy' boundaries between the state and NGOs (Norlund *et al.* 2006: 32; Kerkvliet 2001: 240). As Gainsborough (2010) indicated, a 'key concept in Vietnamese political culture is the idea of the umbrella (*o du*) whereby lower-level institutions or individuals receive backing or protection from those higher up the political chain' (Gainsborough 2010: 57). In the case of the VRN, this backing was provided through VUSTA. As discussed in Chapter 5, the VRN is formally a project of the WARECOD, one of VUSTA's member associations. It is interesting to note that this backing is reflected as part of the NGO's organizational identity. As an illustration, some of the VNGO's websites and staff name cards carry the name of VUSTA along with the name of the NGO (WARECOD 2012).

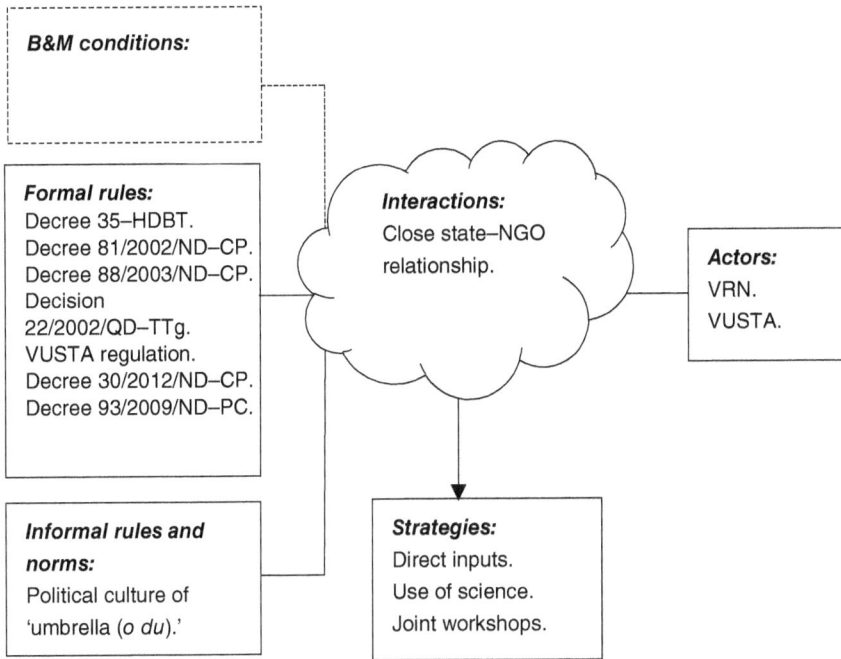

Figure 9.1 Influence of formal rules and actors on the VRN's strategy targeting national decision-makers

There is also a question of whether VUSTA is a state organization or a civil society organization. VUSTA's regulation defines itself as a 'socio-political organisation of the national contingent of science and technique intellects' (Article 1 *VUSTA Regulation* 2006). VUSTA is a member of the Vietnam Fatherland Front (Article 1 *VUSTA Regulation* 2006) and has a close relationship with the Communist Party of Vietnam (CPV) (Norlund *et al.* 2006: 49). Strictly speaking, it is not a government agency. However, its regulation was approved by the Prime Minister (*VUSTA Regulation* 2006). This relationship between VUSTA and the state illustrates the fuzzy boundary between civil society and the state in Vietnam (Kerkvliet 2001: 241; Thayer 2008: 5–11; Norlund *et al.* 2006: 32). VUSTA's close relationship with the CPV, however, does not necessarily cascade down to the associations registered under VUSTA, such as the WARECOD, as they generally have looser connections with the Party (Norlund *et al.* 2006: 49).

Another formal rule that shapes the state–NGO relationship is associated with funding. Many VNGOs and their projects receive funding from international NGOs, and the VRN is no exception (Vietnam Rivers Network 2013: 38). Even when the funding is directly channelled from international NGOs to Vietnamese NGOs, this support needs to be approved by state authorities in accordance with Decree No. 93/2009/ND-CP, which regulates using aid from international NGOs (Article 15 *Regulation on the use of aid from*

INGO 2009). The intervention by the state over the channelling of funds provides room for interference by government agencies in NGOs' activities. This reflects the resource-dependent nature of organizations, including NGOs (Saxon-Harrold 1990; Hsu 2010).

The somewhat 'structured' relationship between the state and NGOs observed in Vietnam is in direct contrast to that of Cambodia, where NGOs also have requirements for registering with the government agencies, specifically the Ministry of the Interior (MoI). However, no specific operational relationship was observed between the MoI and NGOs. NGOs have been operating relatively freely in Cambodia since around 1991 (Cooperation Committee for Cambodia 2012). Unlike in Vietnam, there is no 'umbrella' organization for Cambodian NGOs. This relationship is reflected in the RCC's strategies working with national decision-makers. Compared with the VRN, which used various formal and informal mechanisms for reaching decision-makers, the RCC's work used primarily formal mechanisms. These included formal meetings with CNMC officials during the Xayaburi stakeholder consultation meeting in Sihanoukville, as well as the hydropower dam issues raised during the national workshop on climate change (NGO Forum on Cambodia 2011a; C5 2012a).

One of the formal opportunities to dialogue with the government was created through the PNPCA process of the Xayaburi hydropower dam. The public consultation organized by the CNMC created an opportunity for the RCC to conduct a side meeting with the CNMC during the consultation. The RCC's letter which followed this meeting illustrates the RCC's appreciation

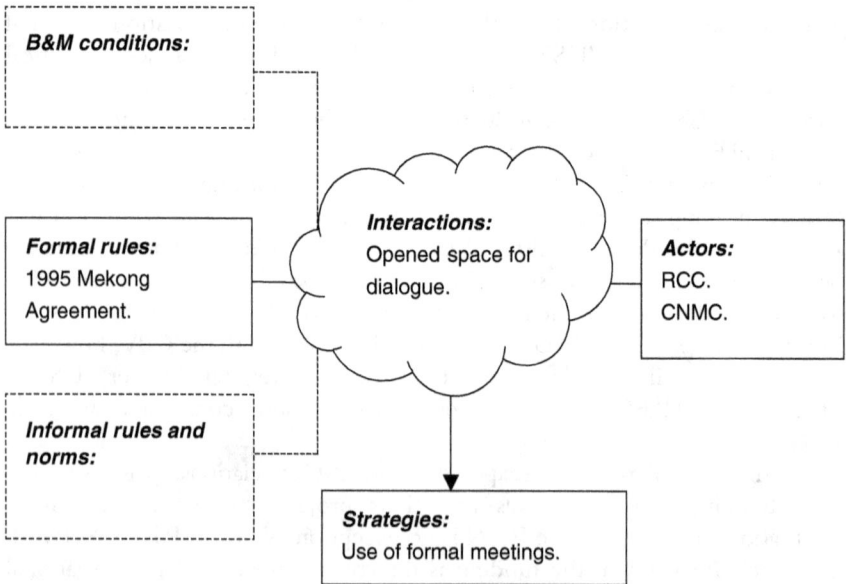

Figure 9.2 Influence of formal rules and actors on the RCC's strategy targeting national decision-makers

of the CNMC's collaborative approach, a positive shift in its relationship with the authority. From this point of view, the Mekong Agreement and the PNPCA, the formal rules governing the Mekong, opened a space for the RCC to dialogue with the national decision-makers, illustrated in Figure 9.2.

It is also important to note that in both the Cambodian and Vietnamese contexts, the position of both governments in association with the Xayaburi dam is that of requesting more assessment of potential impacts from the dam before the decision could be made (Kingdom of Cambodia 2011; Socialist Republic of Viet Nam 2011). This worked favourably to develop a positive relationship between the NGO coalitions and the authorities. In the case of Cambodia, however, when the public consultations were conducted, the official position of the Cambodian government was not yet disclosed. Therefore it is premature to conclude that this position influenced the RCC's strategy of using formal meetings.

Availability of 'credible' science for decision-makers, communicators of science, and cultural and historical aspects of science

A key factor that affected the way both case-study NGO coalitions undertook their strategies was the availability of 'credible' science for decision-makers, and the availability of communicators of science. In Vietnam, science was derived from two sources. One was information from the SEA report, which was commissioned by the MRC and conducted by ICEM. One of the scientists within the VRN's Mekong task force was the team leader of the study in Vietnam (ICEM 2010b). Having a scientist who led the major study commissioned by the MRC gave the VRN the advantage of fully understanding the science behind the study, and gave credibility to its message to the government (V24 2013). Another source of science came from scientists from universities in the Mekong Delta, particularly the Can Tho University. They had been studying ecosystems, soil, and agriculture within the Mekong Delta for decades, and had first-hand knowledge and observation of the changes due to ongoing upstream development occurring within the Delta (V17 2012; Tuan 2011). In both cases, the science came from people and organizations that were perceived as credible by the government. This illustrates the important role of scientists in the VRN and the ways they could develop a science strategy. The importance of the credibility of science communicators is reflected in the literature on the science–policy interface and the way that it is important 'whose science matters' (De Santo 2010).

As discussed above, the VRN had a close working relationship with VUSTA, and VUSTA plays the role of a boundary organization in the Vietnamese context. This role as boundary organization was created by the formal rule that defined VUSTA's role in critically reviewing government policy from a scientific point of view (*Decision No. 22/2002/QD-TTg* 2002; *Decree No. 81/2002/ND-CP* 2002; *Decree No. 88/2003/ND-CP* 2003; *Decree No.30/2012/ND-CP* 2012; *VUSTA Charter* 2012). As discussed earlier, the

VRN's close relationship with the VUSTA and its status as a member association of the VUSTA arises through the formal rules binding Vietnamese NGOs (*Decree No. 35/HTBT* 1992). This close relationship with the organization, which is officially mandated, through formal rules, to provide scientific input to the government's decision-making, allowed the VRN to have wider channels of scientific inputs.

In addition, Vietnam has a culture of respecting science, as discussed in Chapter 6. The importance of science was echoed by some interviewees, who indicated that science and education are important in Vietnam (V11 2012; V16 2012). The interaction of physical and material conditions, formal rules, informal rules and norms, and actors created pathways for a science–policy interface through VUSTA. The interaction also created the close relationship between VUSTA and the VRN. These interactions resulted in allowing the VRN to adopt a strategy of using science. Figure 9.3 illustrates this influence.

In contrast to the situation of the VRN in Vietnam, the RCC in Cambodia did not have its own 'credible' science, or communicators of science. When necessary, the RCC used information from the MRC-commissioned SEA report (C5 2012b). One of the government officials interviewed considered that government agencies have a better understanding of technical information than NGOs, indicating that the RCC members did not provide any added value to the existing knowledge which the government considered it already had (C29 2012).

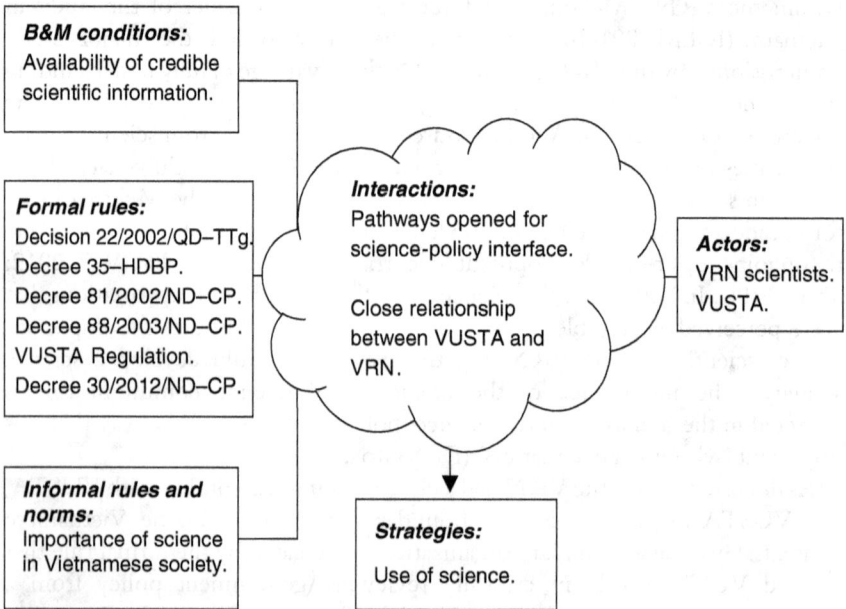

Figure 9.3 Influence of biophysical and material conditions, formal rules, informal rules and norms, and actors on the VRN's use of science

While the RCC attempted to use science in its previous work with the hydropower dams on the Mekong's tributary, the network did not seem to have gained credibility with government officials in its reports. An analysis of what makes 'credible' science and its communicators leads us to consider the influence of the state–NGO relationship. In the case of the RCC, it is important to consider this relationship from an early period when the network was called Se San Working Group, which focused on understanding the impacts of the Yali dam on Cambodian communities (C11 2012; Hirsch and Wyatt 2004). The network conducted and published its own research reports and attempted to communicate with the MRC and the national government authorities (C11 2012; 3S Rivers Protection Network 2007; Rutkow *et al.* 2005; Baird and Mean 2005; Hirsch and Wyatt 2004). The concerns of riparian communities, and the research on impacts on livelihoods supported by the network, were largely neglected by the national agencies that represented Cambodia in international conflict resolution meetings through the MRC (Hirsch and Wyatt 2004: 61–64). Despite the availability of the evidence of impacts on communities, which was based on NGOs' and communities' own research, the Cambodian official representatives reportedly claimed that they did not have adequate 'scientific' evidence to present the case and were thus unable to raise community concerns (Hirsch and Wyatt 2004: 64; C25 2012).

In the case of the Xayaburi dam and the scientific information, some government interviewees commented that Cambodian NGOs tend to raise social human rights issues related to hydropower dams without fully understanding the economics and techniques of the technology (C34 2012; C29 2012). Government agencies consider that they have their own science expertise, and there is a tendency to consider that research outputs produced by Cambodian NGOs are substandard (C7 2012; C32 2012). A Cambodian academic who was formerly engaged in the work of the RCC commented that the RCC tends to use foreign consultants for its scientific work, rather than working with academics within Cambodia (C28 2012). The same interviewee continued to say that even reports prepared by foreign consultants are often not trusted by the government when they are commissioned by Cambodian NGOs (C28 2012).

One of the interviewees commented on the fact that, in comparison with Vietnam, which has a Confucian tradition and places emphasis on the importance of science and education, Cambodia as a society places less emphasis on the use of science as a result of its historical legacy (Horwitz and Calver 1998; V11 2012). Cambodia suffers from a historical loss of intellectuals from the country due to the Khmer Rouge regime, which targeted intellectuals for its genocide. This resulted in the death of many intellectuals and led others to flee the country during the regime (Edwards 2004: 62). The result of this is a large gap in academic knowledge in the country. One interviewee commented that high-level government officials over 50 years of age tend to be former soldiers from the internal conflict period, and do not have an educational basis for understanding science (C3 2012).

In recent years, Cambodian universities have often been considered a place for people to attend in order to receive a degree, rather than an institution for building long-term knowledge through research (C28 2012). The reality of academic life in Cambodia lacks incentive for academics to engage themselves in long-term research (C28 2012; C35 2012). Academics at Cambodian universities are government staff who receive a minimal level of salary, ranging from US$80 to 150 per month on average. This is not a sufficient amount to sustain a life in the capital city, Phnom Penh (C35 2012). In order to sustain their livelihoods, they often engage in external jobs such as working as consultants to donor-funded projects. This leaves them with no extra time to conduct their own academic research (C28 2012; C35 2012). As they are government staff, academics at Cambodian universities also tend to be reluctant to engage with advocacy NGOs that are considered to be anti-governmental (C28 2012). When NGOs need to conduct research on a certain issue, they often contract international consultants who are considered as having higher capacity than Cambodian academics (C28 2012).

The analysis of the importance of the factors discussed above, namely the availability of credible science, the availability of communicators, and state–NGO relationships, is also echoed in the literature of the science–policy interface. A recent study conducted by the Overseas Development Institute, which reviewed the science–policy interface within developing countries, indicates that deliberation between researchers and policy-makers, and advocacy by researchers, were two options favoured by both policy-makers and scientists as measures to promote science, technology, and innovation (Jones *et al.* 2008: 19). The study also identified the existence of a scientific knowledge broker as one of the key important factors in promoting the science–policy interface, and the importance of the credibility of knowledge brokers. The study findings indicate that while professional scientific organizations are considered as most effective mediators between science, technology, and innovation research and policy-making, advocacy organizations are perceived as one of the least effective (Jones *et al.* 2008: 26–31).

This comparison of NGO coalitions illustrates that rules, norms, actors, and physical and material conditions all played important roles in shaping the NGO coalition's advocacy strategies. The Vietnamese case analysis revealed the important role played by boundary organizations in connecting science and policy. The Vietnamese NGOs' access to boundary organizations was shaped through formal rules defining the establishment and registration of NGOs in Vietnam. Credibility for NGOs' science in Vietnam was gained through the characteristics of members of the coalition, and their relationship with the authorities. The Cambodian case lacked boundary organizations and formal rules to support the relationships, and the members did not have particularly strong scientific backgrounds or credibility.

Personal networks and network membership

Another actor that provided the VRN with the avenue to access decision-makers was the VNMC. As discussed in Chapter 5, the VNMC plays the main coordinating role within the Vietnamese government over the issue of the Mekong River (Article 1 *Decision No. 860-TTg* 1995). One of the members of the VRN Mekong task force was a high-ranking official from the VNMC, who was able to play a role of connecting the VRN members effectively with the VNMC (V2 2012; V24 2013).

The analysis of the state–NGO relationship leads us to consider the role that existing networks and social capital played in Vietnam. In Vietnam, the use of informal networks included in the VRN's approach reflects the profiles and human networks brought by the VRN members, particularly among its Mekong task force members. These members include former government officials, scientists at Vietnamese universities, and individuals with networking and facilitation skills (V16 2011; V2 2012).

Some interviewees reflected that personal contacts are very important in Vietnamese society (V2 2012; V24 2013). Referring to this, one interviewee continued that 'who you know is very important in Vietnam' (V24 2013). The author's observation from her personal work experience in Vietnam also supported the view that knowing who you are connected with is very important in gaining trust with Vietnamese people. When the author met Vietnamese people for the first time, they always tried to find out with whom she was connected, and when they found mutual connections the distance between them became less. These anecdotes illustrate the importance of human networks and social capital in Vietnam. This situation is also reflective of the importance of trust within Vietnamese society, as discussed in Chapter 6. Vietnamese traditionally tend to trust small circles of acquaintances through family and village networks.

This cultural context was important for the VRN's advocacy strategy in approaching national decision-makers (V26 2012). In order to gain trust for its advocacy target, the VRN members contacted different people through their networks (V2 2012). The VRN as a network includes government officials, retired government officials, and even an organization which is part of a government institution (V12 2012; Vietnam Rivers Network 2012). This membership characteristic enabled the VRN to develop a good network with government officials. The Vietnamese culture, which emphasizes the importance of personal networks and the need to gain trust through close networks of people, explains the reason for the VRN adopting the strategy of attempting to use all kinds of entry points that were available through the members' networks. As illustrated in Figure 9.4, this cultural aspect, which is also a norm, was one important aspect that shaped the VRN's strategy.

This emphasis on the use of personal networks was not observed in the case of the RCC in Cambodia. The relationship with the government appears to be rather formal, as a key member indicated: 'we write a formal letter requesting

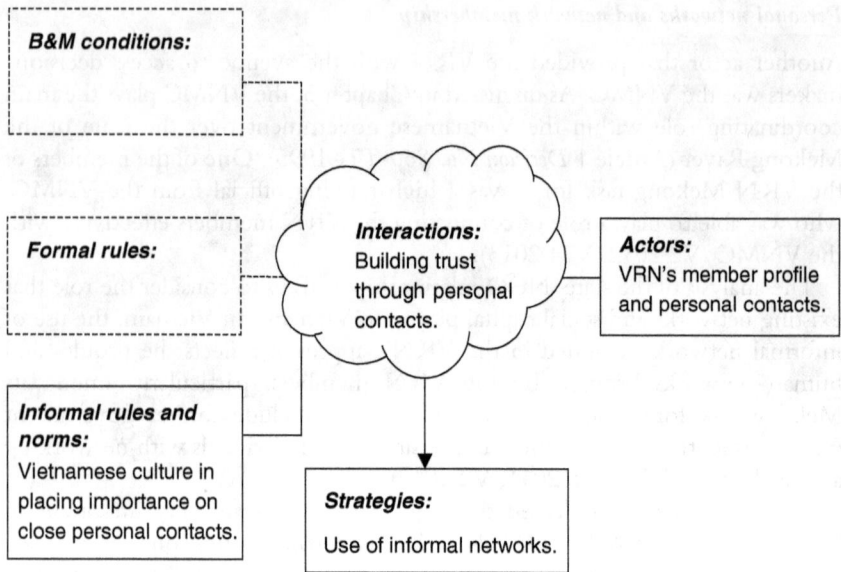

Figure 9.4 Influence of norms and actors on the VRN's use of informal networks

to join any meeting related to the Xayaburi dam' (C5 2012b). The same interviewee indicated that one of the constraints in working with the Cambodian government is the fact that they always need to show respect, and it is difficult to argue with or offer different opinions to the government officials without losing the appearance of respect (C5 2012b; C3 2012). Another interviewee stated that the relationship between NGOs and government in Cambodia is generally not good, as NGOs do not trust the government and, reciprocally, the government may view advocacy NGOs as being anti-government (C28 2012). In addition, as discussed earlier in this chapter, the RCC's relationship in the past with the Cambodian government on the issue of the Se San River was rather contradictory, which may also provide the basis for the current relationship between the RCC and the state.

Building onto existing partnerships

Both the Cambodian and Vietnamese workshops built on existing partnerships and connections with organizations. For the RCC, the workshop at which they collaborated with the Council of Ministers was possible due to the ongoing relationship with the NGO Forum and the CARD (C5 2012b). The experts for the hydropower session were recruited through the RCC's partners internationally (NGO Forum on Cambodia 2011a). Participants in the workshop, particularly the community participants, were available through its members' ongoing relationships with communities (NGO Forum on Cambodia 2011a). While the event was organized jointly with other programmes of the

NGO Forum, in the past the hydropower programme of the NGO Forum and the RCC did not have a particular working relationship with the Council of Ministers (C5 2012b). The RCC took advantage of other programmes within the NGO Forum possessing ongoing working relationships with the Council of Ministers' office, as a way to include hydropower as part of the themes of the public forum (C5 2012b). With its work with the Xayaburi dam, the RCC for the first time appeared to be on good terms with the CNMC, one of its key governmental counterparts (C29 2011).

In the case of Vietnam, VUSTA was the obvious partner for the NGO coalition, as discussed above. In addition, Pan Nature, a Vietnamese environmental NGO, provided an entry point for the VRN to conduct a workshop targeted at members of the National Assembly, using its existing working relationship with the members (V3 2012; V16 2011). Another NGO with which the VRN collaborated in its targeted workshops was WWF. This collaboration started approximately one year after the PNPCA process began (V2 2012). The collaboration with WWF focused on raising awareness and disseminating scientific knowledge on the impacts of upstream hydropower dams, particularly the impacts on sediment flow and geomorphology (WWF and VRN 2011; WWF 2012). For WWF, which has offices in each country within the Lower Mekong Basin, the VRN provided a forum to present its technical knowledge on the impacts of the dam (V8 2012). WWF's approach to the Mekong's mainstream hydropower dams is to be a constructive and solution-oriented partner with governments (Yong and Grundy-Warr 2012: 1045). This aligns with the strategy of the VRN. In conducting the workshop, the WWF provided scientific knowledge and funding to the VRN, and the VRN provided its own scientists and an audience for the WWF's work (WWF and VRN 2011; V8 2012). Although the partnership between the WWF and the VRN is relatively new (V2 2012), there is an alignment in approaches and mutual benefit in its collaboration between these two actors (V8 2012).

These relationships illustrate the resource-dependent nature of partnerships. In particular, this point is highlighted when considering the RCC's relationship with WWF. While WWF operates in both Cambodia and Vietnam, it does not have a working relationship with the RCC. Instead, WWF collaborates with certain members of the RCC on specific projects. WWF has been collaborating with the Cambodian Rural Development Team (CRDT), one of the province-based RCC members, on the conservation of dolphins living in the Mekong River. In line with this work, WWF also supported some of the province-based activities conducted by the CRDT on raising awareness of hydropower dams (C10 2012; C12 2012). WWF also collaborates with Oxfam Australia in supporting the Fisheries Administration in its work analysing the impacts from hydropower dams on livelihoods and nutrition (C22 2012). The collaboration takes place as there appear to be clear mutual benefits in these relationships, which are not observed in the relationship between WWF and the RCC as a network.

Conclusion

Through reviewing the advocacy strategies targeting national decision-makers adopted by the RCC and VRN, this chapter provides an analysis of how rules and norms influence the advocacy strategies of NGO coalitions. Three main findings are observed from the analysis.

First, in both the Cambodian and Vietnamese contexts, formal rules created a situation that influenced state–NGO relationships. The difference lies in the types of rules and the relationships. The relationship created in Cambodia was issue-specific to the Xayaburi dam, where the 1995 Mekong Agreement created a space of dialogue between the RCC and the CNMC over the Mekong River's hydropower dam. In Vietnam, the relationship was created through the national legislation that regulates Vietnamese NGOs. This has the nature of a long-term relationship not specific to the Xayaburi dam.

Domestically, Vietnamese NGOs have to follow tighter regulation than Cambodian NGOs. This tight regulation potentially could be used as a way to control NGOs and civil society movements in Vietnam that threaten the current political regime of the one-party state. As an example, the Vietnamese government does not show tolerance towards dissident movements supported by the Vietnamese people who left the country during the Vietnamese war (Thayer 2008). If these movements become larger, they could potentially threaten one-party rule. The state, therefore, in this situation, may need to carefully control and 'manage' civil society. However, in the case of the VRN, the NGO coalition used the tight regulatory situation to its advantage by creating an entry point in collaboration with state agencies. Both cases illustrate the fact that NGO coalitions utilized the existence of formal rules and their consequences in order to promote and develop their strategies.

The second finding is the role that informal rules and norms played in the development of strategies, particularly in Vietnam. In the VRN's strategy of using informal networks, the Vietnamese norm that places importance on personal networks, and the availability of actors who have effective networks with actors within national government agencies, were both important factors that shaped the strategies of the VRN. In its strategy using science, the norm in Vietnam that values science interacted with three other factors: the close state–NGO relationship shaped through formal rules and its use of actors, the existence of actors and boundary organizations that are capable of communicating science to the decision-makers, and the availability of scientific information.

Finally, the interactions among actors based on resource dependency also contributed to developing strategies. This point echoes the discussion in Chapter 8 analysing the strategies targeting regional decision-makers.

References

3S Rivers Protection Network. 2007. *Abandoned villages along the Sesan River in Ratanakiri Province, Northeastern Cambodia.* Banlung, Ratanakiri, Cambodia: 3S Rivers Protection Network.www.3spn.org/wp-content/uploads/2011/10/3SPN-2007-Abandoned-Villages-Along-the-Sesan-River-Ratanakiri-Province-NE-Cambodia.pdf. Accessed 7 June 2015.

1992 Constitution of the Socialist Republic of Vietnam: As Amended 25 December 2001 (transl. Allens Arthur Robinson). 2001. Vietnam.

Agreement on the Cooperation for the Sustainable Development of the Mekong River Basin. 1995.

Baird, Ian G., and Meach Mean. 2005. *Sesan River Fisheries Monitoring in Rattanakiri Province, Northeast Cambodia: Before and After the Construction of the Yali Falls Dam in the Central Highlands of Viet Nam.* 3S River Protection Network (3SPN), Global Association for People and Environment (GAPE). www.internationalrivers.org/files/attached-files/sesanfisheries2005.pdf. Accessed 13 April 2013.

C3. 2012. Personal interview, 25 July 2012.

C5. 2011. Personal interview, 21 November 2011.

—— 2012a. C5 e-mail communication to the author, 30 August 2012.

—— 2012b. Personal interview, 25 July 2012.

C7. 2012. Personal interview, 27 July 2012.

C10. 2012. Personal interview, 28 July 2012.

C11. 2012. Personal interview, 28 July 2012.

C12. 2012. Personal interview, 30 July 2012.

C13. 2012. Personal interview, 31 July 2012.

C22. 2012. Personal interview, 7 August 2012.

C25. 2012. Personal interview, 8 August 2012.

C28. 2012. Personal interview, 10 August 2012.

C29. 2011. Personal interview, 20 November 2011.

—— 2012. Personal interview, 10 August 2012.

C32. 2012. Personal interview, 16 August 2012.

C34. 2012. Personal interview, 16 August 2012.

C35. 2012. Informal conversations with the author, September–October 2012.

Charter: The Vietnam Union of Science and Technology Associations. 2012. Vietnam.

Cima, Ronald J., ed. 1987. *Vietnam: A Country Study.* Washington, DC: GPO for the Library of Congress.

Cooperation Committee for Cambodia. 2012. *Law on Association and Non-Governmental Organizations (LANGO).* www.ccc-cambodia.org/lango.html. Accessed 16 June 2012; no longer available online.

CARD. undated. *Role and Duties, CARD.* www.card.gov.kh/departments.html. Accessed 18 September 2012; no longer available online.

De Santo, Elizabeth M. 2010. 'Whose science?' Precaution and power-play in European marine environmental decision-making. *Marine Policy* 34 (3): 414–420.

Decision 22/2002/QD-TTg of 30 January 2002 of the Prime Minister on the activities of consultancy, judgment and social expertise by Vietnam Union of Scientific and Technical Associations 2002. Vietnam.

Decision No. 860-TTg of 30 December 1995 of the Prime Minister on the Function, Tasks, Powers and Organization of the Apparatus of the Vietnam Mekong River Committee. 1995. Vietnam.

Decree 35/HTBT of 28 January 1992 of the Council of Ministers on Establishment of non-profit and science and technology organization. 1992. Vietnam.

Decree No. 30/2012/ND-CP of 12 April 2012 of the Government on the organization and operation of social funds and charity funds. 2012. Vietnam.

Decree No. 81/2002/ND-CP of 17 October 2002 of the Government on detailing the implementation of a number of articles of the science and technology law. 2002. Vietnam.

Decree No. 88/2003/ND-CP of 30 July 2003 of the Government on providing for the organization, operation and management of associations. 2003. Vietnam.

Dien, An. 2010. Vietnam lawmakers urge public hearing on Mekong dams. *Thanh Nien News*, 12 November 2010. http://en.baomoi.com/Home/sciencetechnology/thanhniennews.com/Vietnam-lawmakers-urge-public-hearing-on-Mekong-dams/84846.epi. Accessed 18 June 2012.

—— 2012. Work has begun on Xayaburi dam, Thai firm says. *Thanh Nien News*, 27 April. www.thanhniennews.com/2010/Pages/20120427-Work-has-begun-on-Xayaburi-dam-Thai-firm-says.aspx. Accessed 9 February 2013.

Edwards, Matthew. 2004. The rise of the Khmer Rouge in Cambodia: internal or external origins? *Asian Affairs* 35 (1): 56–67.

Gainsborough, Martin. 2010. *Vietnam: Rethinking the State.* London: Zed Books.

Government of Lao PDR, and Pöyry. 2011. *Compliance Report: Xayaburi Hydroelectric Power Project. Run-of-River Plant.* www.poweringprogress.org/download/Reports/2012/July/Compliance%20Report%20Xayaburi%20Main%20Final.pdf. Accessed 7 June 2015.

Hirsch, Philip, and Andrew Wyatt. 2004. Negotiating local livelihoods: scales of conflict in the Se San River Basin. *Asia Pacific Viewpoint* 45 (1): 51–68.

Hirsch, Philip, Kurt Mørck Jensen, Ben Boer, Naomi Carrard, Stephen FitzGerald, and Rosemary Lyster. 2006. *National Interests and Transboundary Water Governance in the Mekong.* http://sydney.edu.au/mekong/documents/mekwatgov_mainreport.pdf. Accessed 23 July 2011.

Horwitz, Pierre, and Michael Calver. 1998. Credible science? Evaluating the regional forest agreement process in Western Australia. *Australasian Journal of Environmental Management* 5 (4): 213–225.

Hsu, Carolyn. 2010. Beyond civil society: an organizational perspective on state–NGO relations in the People's Republic of China. *Journal of Civil Society* 6 (3): 259–277.

Huynh, Tran Thu, and Dau Anh Tuan. 2007. *Vietnamese Business Associations as Policy Advocates: A Lot More Can Still be Done.* GTZ and VCCI. www.sme-gtz.org.vn/Portals/0/AnPham/Bussiness%20Associations%20as%20Policy%20Advocates.pdf. Accessed 7 June 2015.

ICEM. 2010a. *MRC Strategic Environmental Assessment (SEA) of Hydropower on the Mekong Mainstream: Summary of the Final Report.* Hanoi. www.mrcmekong.org/assets/Publications/Consultations/SEA-Hydropower/SEA-FR-summary-13oct.pdf. Accessed 7 June 2015.

—— 2010b. *MRC Strategic Environmental Assessment (SEA) of the Hydropower on the Mekong Mainstream.* Hanoi: Mekong River Commission. www.mrcmekong.org/assets/Publications/Consultations/SEA-Hydropower/SEA-Main-Final-Report.pdf. Accessed 3 May 2014.

International Rivers, 3S River Protection Network, My Village, Cambodian Rural Development Team, CD Cam, FACT, and NGO Forum on Cambodia. 2011. *The Xayaburi Dam: A Looming Threat to the Mekong River.* Xayaburi dam fact sheet (in Khmer). www.internationalrivers.org/files/attached-files/xayaburi_fact_sheet_kh.pdf. Accessed 15 September 2012.

Jones, Nicola, Harry Jones, and Cora Walsh. 2008. *Political Science? Strengthening Science–Policy Dialogue In Developing Countries*. Working Paper 294. www.odi.org.uk/sites/odi.org.uk/files/odi-assets/publications-opinion-files/474.pdf. Accessed 27 April 2013.

Katsumata, H. 2003. Reconstruction of diplomatic norms in Southeast Asia: the case for strict adherence to the 'ASEAN Way'. *Contemporary Southeast Asia* 25 (1): 104–121.

Kerkvliet, Benedict J. Tria. 2001. An approach for analysing state-society relations in Vietnam. *Sojourn: Journal of Social Issues in Southeast Asia* 16 (2): 238–278.

Kingdom of Cambodia. 2011. *Mekong River Commission Procedures for Notification, Prior Consultation and Agreement: Form/Format for Reply to Prior Consultation*. Cambodia National Mekong Committee (CNMC). www.mrcmekong.org/assets/Consultations/2010-Xayaburi/Cambodia-Reply-Form.pdf. Accessed 25 November 2011.

MRC. 2011. *Prior Consultation Project Review Report: Volume 2 Stakeholder Consultations related to the proposed Xayaburi dam project*. Mekong River Commission. www.mrcmekong.org/pnpca/2011-03-31-Report-on-Stakeholder-Cons-on-Xayaburi.pdf. Accessed 11 April 2011.

—— undated. *Strategic Environmental Assessment of Mainstream Dams*. Mekong River Commission. www.mrcmekong.org/about-mrc/programmes/initiative-on-sustainable-hydropower/strategic-environmental-assessment-of-mainstream-dams. Accessed 12 April 2013.

NGO Forum on Cambodia 2011a. Minutes, National Conference on Climate Change, Agriculture and Energy, Phnom Penh, 1-2 December 2011.

—— 2011b. *Thank you letter for your kind collaboration with Rivers Coalition in Cambodia and civil society concerns on the proposed Xayaburi dam*. A letter to H.E. Lim Kean Hor, Chairman of Cambodia National Mekong Committee, 23 March 2011.

Norlund, Irene. 2007. *Filling the Gap: The Emerging Civil Society in Vietnam*. Ha Noi: UN Vietnam. www.un.org.vn/en/component/docman/doc_details/3-filling-the-gap-the-emerging-civil-society-in-viet-nam.html. Accessed 14 April 2012.

Norlund, Irene, Dang Ngoc Dinh, Bach Tan Sinh, Dang Ngoc Quang, Do Bich Diem, Nguyen Manh Cuong, Tang The Cuong, and Vu Chi Mai. 2006. *The Emerging Civil Society: An Initial Assessment of Civil Society in Vietnam*. Hanoi: Vietnam Institute of Development Studies (VIDS), UNDP Vietnam, and SNV Vietnam. www.civicus.org/new/media/CSI_Vietnam_report%20.pdf. Accessed 7 December 2011.

Quới, Lê Phát 2011. Quan hệ dòng chính sông Mekong với đồng bằng sông Cửu long [*The relationship between the Mekong River's mainstream with Cửu long Delta*]. Presentation at Open Dialogue: Xayaburi dam and hydropower cascade development in the mainstream of the Lower Mekong river basin. GAINS–LOSSES to Vietnam and downstream countries. Hanoi, 15 March 2011.

R2. 2011. Personal interview, 16 November 2011.

Regulation on management and use of foreign nongovernmental aid promulgated by Decree No.93/2009/ND-CP of 22 October 2009 of the Government 2009. Vietnam.

Regulations of Vietnam Union of Science and Technology Associations. Issued jointly with the decision No 650/QD-TTg, 24 April 2006 of the Prime Minister of the Socialist Republic of Vietnam 'Approval of the Regulations of Vietnam Union of S-T Associations' 2006. Vietnam.

Rutkow, Eric, Cori Crider, Tyler Giannini, and Allison Friedman. 2005. Down River: The Consequences of Vietnam's Se San River Dams on Life in Cambodia and Their Meaning in International Law. NGO Forum on Cambodia. http://hrp.law.harvard.edu/wp-content/uploads/2013/02/Down-River-2005.pdf. Accessed 7 June 2015.

Saxon-Harrold, Susan K. 1990. Competition, resources, and strategy in the British nonprofit sector. In *The Third Sector: Comparative Studies of Nonprofit Organizations*, edited by H. Anheier and W. Seibel. Berlin: Walter de Gruyter & Co.

Socialist Republic of Viet Nam. 2011. *Mekong River Commission Procedures for Notification, Prior Consultation and Agreement Form for Reply to Prior Consultation*. The Viet Nam National Mekong Committee. www.mrcmekong.org/assets/Consultations/2010-Xayaburi/Viet-Nam-Reply-Form.pdf. Accessed 25 October 2011.

TERRA. 2008. *Watershed: People's Forum on Ecology*. Burma, Cambodia, Lao PDR, Thailand, Vietnam, Towards Ecological Recovery and Regional Alliance (TERRA). www.terraper.org/mainpage/watershed.php. Accessed 2 November 2011.

Thanh Nien News. 2012. Vietnam scientists ask PM to stop Lao dam project. *Thanh Nien News*, 10 May. www.thanhniennews.com/2010/pages/20120510-vietnamese-experts-seek-pm-intervention-to-stop-mekong-dam-project.aspx. Accessed 4 June 2012.

Thayer, C.A. 2008. One party rule and the challenge of civil society in Vietnam. In *Remaking the Vietnamese State: Implications for Vietnam and the Region*, Vietnam Workshop, Hong Kong, 21–22 August 2008.

Thien, Nguyen Huu. 2012. Nguy cơ tác động của thủy điện dòng chính Hạ lưu vực Mekong đến ĐBSCL [*Potential impacts of the Mekong River's mainstream hydropower dams on the Cuu Long Delta*]. Presentation at Mekong and hydropower dams workshop, Ho Chi Minh City, 14 August 2012.

Trandem, Ame. 2011. *Milestones of Concern: A Timeline of Concerns Expressed over the Proposed Xayaburi Dam*. September 2008 to July 2011. (Document obtained from the author.)

Tuan, Le Anh. 2010. Thách thức của phát triển thuỷ điện dòng chính sông Mekong đến Đồng bằng sông Cửu Long: Phù sa và Hệ thống nông nghiệp [*Development of hydropower in the mainstream of the Mekong River and challenges to the Cuu Long River Delta: sediment and agricultural system*]. Presentation at Policy Dialogue: Hydropower dam development in the Mekong River and the challenges for Vietnam, Hanoi, 7 November 2010.

—— 2011. Nguồn nước sông Mekong và vấn để an ninh lương thực ở đồng bằng sông Cửu Long: Rủi ro và thách thức [*Mekong Water Resources and Food Security in the Cuu Long River Delta. The Risk and Challenges*]. Presentation material provided by the author.

V2. 2011. Personal interview, 23 November 2011.

—— 2012. Personal interview, 10 July 2012.

V3. 2012. Personal interview, 10 July 2012.

V7. 2012. Personal interview, 11 July 2012.

V8. 2012. Personal interview, 13 July 2012.

V10. 2012. Personal interview, 14 July 2012.

V11. 2012. Personal interview, 14 July 2012.

V12. 2012. Personal interview, 16 July 2012.

V16. 2011. Personal interview, 23 November 2011.

—— 2012. Personal interview, 19 July 2012.

V17. 2012. Personal interview, 19 July 2012.

V20. 2011. Personal interview, 1 December 2011.

V24. 2013. Informal conversations with the author, 16–21 June 2013.

V25. 2012. Personal interview, 19 July 2012.

V26. 2012. Personal interview, 10 July 2012.

Vietnam Rivers Network. 2011a. Nhận xét về báo cáo của Pöyry [*Comments on Pöyry report by the VRN*]. (Document obtained from an interviewee.)

—— 2011b. Workshop materials from Training Workshop on Coastal Geomorphology and Sediment Transit, Ho Chi Minh City, 22 November 2011.

—— 2012. About Us/Introduction. http://vrn.org.vn/en/h/d/2012/04/244/Introduc tion/index.html. Accessed 7 November 2012.

—— 2013. Báo Cáo Tông Kêt Hoạt Dông (Activity Report) 2006–2012. http://vrn.org. vn/media/files/VRN%20annual%20report_Final.pdf. Accessed 12 February 2013.

VUSTA, WARECOD, VRN, VNWP, CIWAREM, and CEWAREC. 2011. Press release: Open Dialogue. Xayaburi dam and hydropower cascade development in the mainstream of the Lower Mekong river basin: GAINS-LOSSES to Vietnam and downstream countries.

WARECOD. 2011. Letter of petition. A letter addressed to Mr Nguyen Tan Dzung Prime Minister; Government Office; Ministry of Natural Resources and Environment; Ministry of Industry and Commerce; Ministry of Agriculture and Rural Development; Vietnam Mekong River Committee; Vietnam Union of Technology and Science Associations, 15 April 2011.

—— 2012. Who are we? www.warecod.org.vn/en/thong-tin/about-us/52.aspx. Accessed 23 June 2013.

Wells-Dang, Andrew. 2014. Civil society networks in Cambodia and Vietnam: a comparative analysis. In *Southeast Asia and the Civil Society Gaze: Scoping a Contested Concept in Cambodia and Vietnam*, edited by G. Waibel, J. Ehlert, and H. N. Feuer. New York: Routledge.

WWF. 2012. Workshop presentations: Knowledge of sediment transport and discharges in relation to fluvial geomorphology for detecting the impact of large-scale hydropower projects in the Mekong River basin. http://wwf.panda.org/what_we_do/where_we_ work/greatermekong/?205051/WWF---MRC-Workshop. Accessed 31 October 2012.

WWF, and VRN. 2011. Agenda: Training Workshop on Coastal and Geomorphology and Sediment Transit, Ho Chi Minh City, 22 November 2011.

Yong, Ming Li, and Carl Grundy-Warr. 2012. Tangled nets of discourse and turbines of development: Lower Mekong mainstream dam debates. *Third World Quarterly* 33 (6): 1037–1058.

10 Strategies targeting stakeholders in affected areas

Introduction

The Xayaburi hydropower dam is expected to significantly impact downstream areas in Cambodia and Vietnam. As discussed in Chapter 4, the direct impact in Cambodia is expected in areas along the mainstream of the Mekong River. In Vietnam, most impact is expected to occur in the Mekong Delta. Both the Rivers Coalition in Cambodia (RCC) and the Vietnam Rivers Network (VRN) conducted awareness-raising activities for communities and stakeholders living within the potentially affected areas. This chapter first discusses the strategies adopted by the RCC, followed by a discussion of VRN's strategies. The chapter is followed by a comparative analysis of the strategies and the recurring themes identified within them, with the aim of identifying how rules and norms influenced the advocacy strategies of the case-study coalitions.

Strategies adopted by the RCC

Community awareness-raising events

RCC members organized various community-related events during the Xayaburi Procedures for Notification, Prior Consultation and Agreement (PNPCA) discussion in order to raise awareness of the Xayaburi dam. In addition, thumb print petitions by villagers, which targeted regional decision-makers (discussed in Chapter 8), also functioned as awareness-raising exercises among community members. Events took place at various localities within Cambodia. This section discusses the key events that took place in Kratie and Kampong Cham provinces, as well as activities in other localities.

Community events in Kratie

Kratie province is situated along the Mekong River's mainstream (see Figure 4.2) and faces significant threats from the development of hydropower dams on both the mainstream and tributaries. Two of the hydropower dams planned on the mainstream of the Mekong River in Cambodia, namely Sambor dam in Kratie province and Stung Treng dam in Stung Treng province, are in the vicinity of Kratie province. When built, these planned dams are expected to result in significant impacts in the area (ICEM 2010). According to the strategic environmental assessment (SEA) of the Mekong's mainstream dams, these two dams could effectively stop fish migration from Cambodia to Laos, which is the key migration route of fish species in the Mekong River (ICEM 2010: 95). This situation is important as these plans for dam construction by the Cambodian government may have potentially affected the attitude of the local authorities in Stung Treng and Kratie provinces related to the hydropower dams.

Two Kratie-based RCC member NGOs were active in organizing community events: the Cambodian Rural Development Team (CRD) and the Community Economic Development (CED). Both organizations work with community livelihoods improvement, primarily in Kratie and Stung Treng provinces (Cambodian Rural Development Team undated; C13 2012). At times, these NGOs talked about the impact of hydropower dams on the Mekong River during their regular visits to communities and through their rural development project activities (C12 2012; C13 2012). In June 2012 the CRDT and the CED organized two awareness-raising events. The first was a community awareness-raising event in Sambor district, Kratie province on 20 June 2012 (C12 2012). The event involved presentations on natural resource use given by the international and local NGOs working on natural resources issues along the Mekong mainstream. These were followed by a peaceful march (C12 2012; C13 2012). The event attracted approximately 130 participants from the community, including village chiefs, commune chiefs, local government officers, and local NGOs. Another event was organized on 28 June 2012, this time held at the University of Management and Economy in Kratie province (C12 2012). The event was attended by university students as well as by commune leaders in the area (C12 2012). In a similar way as in the earlier event, international and local NGOs working in the area, including the World Wide Fund for Nature (WWF), Oxfam, and the CRDT, presented the results of their research on conservation, climate change, disaster, and livelihoods that are linked to the Mekong River, particularly related to Kratie and Stung Treng provinces (C12 2012). After the presentations, students wrote their ideas and comments about the topics on a big banner (C12 2012).

The organizers' initial intention for both events was to raise local awareness on the Xayaburi dam. However, after reluctance from the local authorities to place hydropower as the topic of the event, the wording describing the purpose of the event was formulated as to 'raise awareness on the importance of protecting the benefits from the Mekong River' (C13 2012; C12 2012; C10 2012). This incident exemplifies the influence of the informal pressure from local authorities which the RCC faced in organizing the event.

Both events were funded by International Rivers (IR), which also supported the provincial NGO network where RCC members shared information related to the development of hydropower dams in the region with other NGOs operating in Kratie province (C12 2012; C13 2012). According to one interviewee, the details of activities funded by IR were determined through joint discussions between the donor and the recipient of the fund (C13 2012). This illustrates the fostering relationship between IR and the RCC members.

Buddhist peace walk in Kampong Cham

Kampong Cham is located downstream from Kratie province on the mainstream of the Mekong River (see Figure 4.2). The Buddhist Association for Environmental Development (BAED), a relatively new member of the RCC since the beginning of 2012, is based in this province (C9 2012b). The BAED organized a peace walk campaign to stop the Xayaburi dam on 29 June 2012 (NGO Forum on Cambodia 2012a). The purpose of this event was to raise awareness on the Xayaburi hydropower dam among communities living along the Mekong River, and to gain their support for cancellation of the Xayaburi dam (C9 2012b). The event also provided an opportunity for the communities to raise any concerns about the Xayaburi dam (C9 2012b). During this event, approximately 500–600 participants marched in the town of Kampong Cham, a provincial town along the Mekong River (see Figure 4.2) (Lipes 2012; C9 2012b). Participants included monks, local communities, local authorities, students, and communities from other parts of Cambodia such as Rattanakiri, Stung Treng, Mondulkiri, and Kandal provinces (C9 2012b; C17 2012). This spread of participants illustrates the reach of the RCC's network. The event was also attended by the Deputy Secretary General of the Cambodian National Mekong Committee (CNMC) (C3 2012; C29 2012; Lipes 2012). The participation of a high-ranking official from the CNMC highlighted the support by the Cambodian government for the initiative. The event was funded by IR (C3 2012; C9 2012b). The idea for the march emerged through discussions between IR and the RCC members, including the BAED who implemented the event (C9 2012b; C3 2012).

Prior to the peace walk, the RCC conducted activities in seven districts to inform communities about hydropower dam impacts and the Xayaburi dam (C9 2012b). Before these meetings, RCC members provided training for Buddhist monks on the Xayaburi dam (24–25 January 2012). This allowed Buddhist monks to become communicators to the villagers on the Xayaburi issues (C9 2012b; C23 2012). The NGO sector review by the Cooperation Committee for Cambodia (CCC) indicates that advocacy through Buddhist-led community groups is a relatively new form of advocacy, which is more acceptable to the government compared with the more direct approaches introduced by foreign NGOs in the 1990s. These are still perceived as confrontational (Bañez-Ockelford 2010: 26). However, one interviewee commented that the degree to which Buddhist monks can play a role is limited to acting as communicators (C8 2012). This limited role of Buddhist monks is related to the current political climate of Cambodia and originates from the suppression of peaceful demonstrations organized by Buddhist monks in 1998. The RCC members involved in organizing the event commented that they initially faced resistance by provincial government officials, but the officials eventually approved the event (C23 2012; C9 2012a).

Other community-related events and activities

In addition to these community events, the RCC member NGOs also discussed and disseminated information on the Xayaburi hydropower dam to the communities through their existing community networks (C12 2012). As an example, the Fisheries Action Coalition Team (FACT), an RCC member NGO consisting primarily of a network of fishermen in Cambodia, disseminated information about the Xayaburi hydropower dam directly to fishermen living in other parts of Cambodia (C6 2012). Other RCC members also disseminated information about the dam to villagers on various occasions (C13 2012; C18 2012; C12 2012). At times, NGO staff faced a situation where the village chief or the commune chief was not happy for NGOs to discuss dam issues with the villagers, as local leaders do not want villagers to be against the dam development (C12 2012).

The network of RCC members facilitated collaboration among communities within Cambodia that would not have been possible otherwise. As an example, Rattanakiri is a province upstream of the Mekong River (see Figure 4.2), thus its communities will not experience any direct impact from the Xayaburi dam. However, the communities living along the tributaries of the Mekong River in this area have already been impacted by other hydropower dams through changes in flood patterns and in water quality and flow which have subsequently affected natural resources, particularly fisheries, and the health of local residents (Wyatt and Baird 2007). The first-hand experience of these communities of the impacts of dams made them sympathetic to the potential impact from the Xayaburi hydropower dam on the downstream communities (C17 2012), and they collaborated with the Xayaburi advocacy work led by the RCC. Representatives of these communities participated in the peace walk in Kampong Cham and presented their experiences from other hydropower dams to participants at the event (C17 2012).

Thumb print petitions by community members

Presenting a petition to authorities using thumb prints was another tactic used by the RCC. However, this activity was also considered 'sensitive' by the authorities, and this influenced the RCC's activities (C12 2012; C13 2012; C16 2012). As discussed in Chapter 8, the RCC sent a petition letter to the Thai Prime Minister along with thumb prints of 3,208 Cambodian community representatives in December 2011, demanding that the Thai government stop its plans to buy electricity from the Xayaburi dam (NGO Forum on Cambodia 2012b). Communities living along the tributaries of the Mekong that would not be directly affected by the Xayaburi dam also collaborated in this petition, as 896 community members in this region thumb printed the petition to cancel the Xayaburi hydropower dam which was addressed to the Prime Minister of Thailand (3S Rivers Protection Network 2011: 9).

During the Buddhist peace walk organized in Kampong Cham province on 29 June 2012 (discussed above), 186 members of communities, monks, students, and CSOs signed another petition letter demanding a stop to the Xayaburi dam (NGO Forum on Cambodia 2012b). Some of these petitions and the thumb prints were displayed to the public at one of the national-level forums organized by the RCC (C16 2012). According to one RCC member, a high-ranking government official who attended the forum saw this petition and became angry, criticizing the organizer for displaying it (C16 2012: 8). The interviewee commented that this reaction was due to the fact that collecting thumb prints is considered to be an action 'against the government' (C16 2012: 9), further illustrating the negative perception by government officials of thumb prints.

Collecting thumb prints was also considered to be a 'sensitive' activity by some NGO members. Although thumb print collection was an agreed activity among the RCC members, one interviewee commented that there was disagreement among the RCC members on this activity (C16 2012). Some of the member NGOs did not participate in the thumb print collection activity as they were afraid it would be considered too controversial in the eyes of local governments, and they did not want to jeopardize their carefully cultivated relationships with the local authorities (C16 2012; C15 2012). Another member NGO initially tried to collect thumb prints directly from community members until the local authority opposed this activity (C13 2012). Therefore the NGO changed its strategy, and instead of collecting thumb prints directly from community members, left the petition form with the community, leaving them to collect the thumb prints themselves (C13 2012).

For many community members, this activity was also considered to be sensitive. Some of the communities faced difficulties in collecting thumb prints when local authorities such as commune councils described the activity as 'against the government's policy', which is pro-development (C12 2012). This situation implies that anti-dam activities were considered anti-development by the government. An RCC member interviewee observed that some local authorities thought that opposing the Xayaburi dam could result in opposition to hydropower dams planned in Cambodia, which might indicate that the communities were against the government (C12 2012). Some community members were afraid when the village head told them that they would get into trouble if they were against the dam, even though it is not clear what kind of 'trouble' or 'punishment' they might encounter (C12 2012).

As an example of possible 'trouble', a man in Phnom Penh was detained by police while collecting thumb prints from villagers for a petition demanding the release of Mam Sonando, an independent radio station owner, who was arrested for political dissent (Sovuthy 2012). (The case of Mam Sonando is discussed more in detail in Chapter 11.) Other villagers were not afraid of getting into any trouble (C12 2012; C17 2012). Some villagers expressed their scepticism about thumb print activities as they were afraid that the collected thumb prints might be misused by anyone who might receive a copy of the thumb prints (C20 2012). While there were different opinions and perceptions of this activity, it was clear that there was informal pressure from the local authorities against thumb print collection activities in general.

Strategies adopted by the VRN

In the Mekong Delta in Vietnam, the VRN organized a series of dialogue workshops to discuss hydropower dams with the local stakeholders in the area (V10 2012). These workshops were one of the first attempts by the VRN to work with local stakeholders within the Mekong Delta on the issues of hydropower dams (V16 2011). They took place almost one year after the PNPCA process commenced, illustrating the fact the VRN's initial focus was to interact directly with the national-level decision-makers. One interviewee indicated that 'we realize that out of 18 million people in the Delta, most people don't know how they would be affected' (V16 2011).

A series of three workshops were held initially, targeting various stakeholder groups from government officials to local farmers. These workshops were conducted in collaboration between the VRN and Can Tho University, to which some of the scientists from the VRN's Mekong task force also belonged (V16 2012). The first workshop was held on 28 July 2011 in Can Tho city, and targeted provincial leaders and members of staff from the provincial departments (V16 2012; V10 2012; V17 2012). The second workshop was held on 26–27 November 2011, just before the MRC council meeting in December 2012 when the ministers from the four riparian countries were supposed to make a decision on the Xayaburi dam. The dialogue discussed the impacts of the upstream dam on people's livelihoods and was targeted at farmers within the Mekong Delta (V16 2012; V10 2012; V17 2012). Approximately 50 farmers from eight provinces participated in the workshop. Farmers from the other provinces were unable to attend due to the floods (Ni 2012; V16 2012). The farmers' unions from each province mobilized participants (V16 2012). The third dialogue workshop was held on 12 July 2012, targeting provincial leaders, farmers, and scientists (V16 2012; V10 2012; V17 2012).

Another workshop was held within the Mekong Delta, specifically targeting provincial government officials (V16 2012). The workshop was organized in collaboration with the Southwest Steering Committee, a politically important committee chaired by the Vice Minister and located directly under the bureau within the central Party in Hanoi (V15 2012). On 25 November 2011, the VRN organized a workshop co-chaired by the Southwest Steering Committee and the Research Centre of Forest and Wetlands (ForWet), a Vietnamese NGO based in Ho Chi Minh City (V16 2011). The workshop targeted the provincial governments. The Southwest Steering Committee is a governmental institution consisting of 13 southern provinces in Vietnam with the mandate to consult with southern provinces on issues related to policy and strategy in the Mekong Delta (V15 2012). As commented by one interviewee, 'the most important outcome [from the workshop] was a letter from the Southwest Committee to the central government agencies' (V16 2012; Southwest Steering Committee 2011).

Comparative analysis

The RCC used different mechanisms to raise the awareness of the communities on the Mekong River, including workshops, thumb print petitions, community-level events, and meetings. Events were organized at different localities by the RCC member NGOs located in the provinces of Stung Treng, Kratie, Kampong Cham, and Rattanakiri (C18 2012; C14 2012; C5 2012; C12 2012; C9 2012b). In contrast, the awareness-raising activities of the VRN in the Mekong Delta during the study period were limited to three workshops. Cambodian events were held in communities or provincial towns, allowing participation by various individuals from within the communities, including farmers, government officers, and students. In contrast to Cambodia, where most events were attended by local participants, most participants in the Vietnamese workshops travelled from other provinces to the events. This made it logistically difficult for other community members to participate. In Cambodia, different events were held at local level throughout the PNPCA process of the Xayaburi dam, aiming to raise awareness of the impacts of hydropower dams on the Mekong River. Compared with Cambodia, where local-level activities took place throughout the PNPCA process, the VRN's activities at local level did not take place until one year after the PNPCA process commenced.

When organizing the events, the RCC member NGOs often faced resistance from local authorities towards raising the issue of hydropower dams, resulting in modifications as to how events were presented. At times, the name of the event had to be changed from 'hydropower' to 'raising awareness on the importance of protecting the Mekong River' (C13 2012; C10 2012; C12 2012). As discussed in Chapter 9, the RCC was also hesitant to raise hydropower as a main issue of discussion at the national-level workshop, and used the wording 'energy' instead (C5 2012). In contrast, the workshops in Vietnam clearly indicated hydropower as a main topic, and none of the interviewees in Vietnam mentioned the existence of informal pressure from the authorities against raising hydropower dam issues openly at the provincial level. As discussed in Chapter 9, the VRN's national-level workshops also clearly identified hydropower as a main topic of discussion.

Where did the differences in approach come from? Through a comparative analysis of the two case studies, two themes emerge: the authorities' tolerance over advocacy on sensitive issues; and NGO–community relationships.

Authorities' tolerance over advocacy on sensitive issues

The comparison of Cambodia and Vietnam highlights the difference in sensitivity of the respective authorities towards discussing the issue of hydropower dams. As discussed in Chapter 9 and here, the RCC faced informal pressure by national and local authorities not to raise the issue of hydropower dams. On the other hand, none of the interviewees in Vietnam indicated the existence of informal pressure against raising the issue of hydropower dams at the workshops that the VRN conducted in the Mekong Delta. The VRN members were also able to openly and publicly discuss the costs and benefits of hydropower dams on the Mekong with the National Assembly members and the Vietnamese government (Vietnam Rivers Network 2010; VUSTA *et al.* 2011). This situation in Vietnam may be also due to the Vietnamese government's position over the Mekong's mainstream dams. Vietnam does not have a strong interest in investing in mainstream dams on the Mekong, as discussed in Chapter 4. In addition, the Vietnamese Prime Minister is from the Mekong Delta, and had a good understanding of the potential impact of the Mekong's hydropower dams on the Delta (V24 2013; V2 2012).

The impact of hydropower dams in general has been recognized for some time in Vietnam, and according to a comment by one interviewee, the government appears to be open to criticism:

> We did have a lot of strong critiques in Vietnam already. The hydropower issue has been discussed in the National Assembly. The government also recognized there is a problem and they are trying to fix it.
>
> (V20 2011)

The fact that the VUSTA was engaged in investigating the impacts from hydropower dams in Vietnam may have helped secure the government's acceptance of debates about hydropower impacts. As an example, the Hoa Binh Union of Science and Technology Association, which is the provincial branch of VUSTA, had been studying the impacts of resettled and downstream communities from the Hoa Binh hydropower dam. The Hoa Binh hydropower plant was built in the 1980s and started operation in 1994 (Phuc 2011). In another case, VUSTA and its associated scientists were also conducting a study on the impacts of the Thac Ba hydropower dam, built over 30 years ago, and on the impacts of resettlement of communities as a result of the Son La hydropower dam, whose construction was still on going (VUSTA 2007, 2006).

Collaboration with the VUSTA was identified as an important strategy by the VRN. This collaboration supported the VRN in its efforts to raise a sensitive subject. As one key VRN member commented,

> You know when you want to do something sensitive, you have to find the way, find people. But you have to talk in a way to make it less controversial, and less conflict. You have to talk constructively, so people see you in that way. You can be in trouble, then if you are in trouble, you cannot help other people. That's why we want to work with VUSTA from the beginning.
>
> (V10 2012)

However, there were limitations in the NGO coalition's capacity to raise the issue of hydropower development with the government. Although the issue of the Mekong's mainstream hydropower dams was not a sensitive one for the government, one interviewee indicated that the issue of dams on the tributaries of the Mekong was not allowed to be raised during one of the workshops which the VRN organized jointly with VUSTA (R8 2012). This situation is associated with the fact that while Vietnam only has one project proposal on the mainstream of the Mekong River, it has several ongoing and planned projects on the tributaries of the Mekong River which will affect downstream communities in Cambodia (Grimsditch 2012). In addition, Electricity Vietnam (EVN), the state-owned electricity company in Vietnam, invested in the construction of the Lower Se San 2 dam, which is located within the Cambodian section of the Se San River, a tributary of the Mekong (Grimsditch 2012).

In Vietnam, the government does not tolerate citizens' actions that directly threaten the rule of the Communist Party (Sidel 1997: 289; Thayer 2008). However, the government appears to show tolerance over other issues. For instance, a CIVICUS (World Alliance for Citizen Participation) report indicates that the government accepts advocacy on issues such as development and social programmes or humanitarian relief programmes (Norlund *et al.* 2006: 73). Over the issues associated with the Xayaburi dam, one interviewee commented that 'the political space in Vietnam goes up and down, depending on the political situation' (R8 2012).

As discussed in Chapter 6, 'fence-breaking' (*pha rao*) is a norm that exists in Vietnamese society. This can also provide an explanation for the government's tolerance over certain forms of criticism of its policy. This cultural norm provides an explanation of the way that the state accepts certain requests from civil society, and this may have played a role in the case of the VRN's strategy to openly discuss hydropower issues.

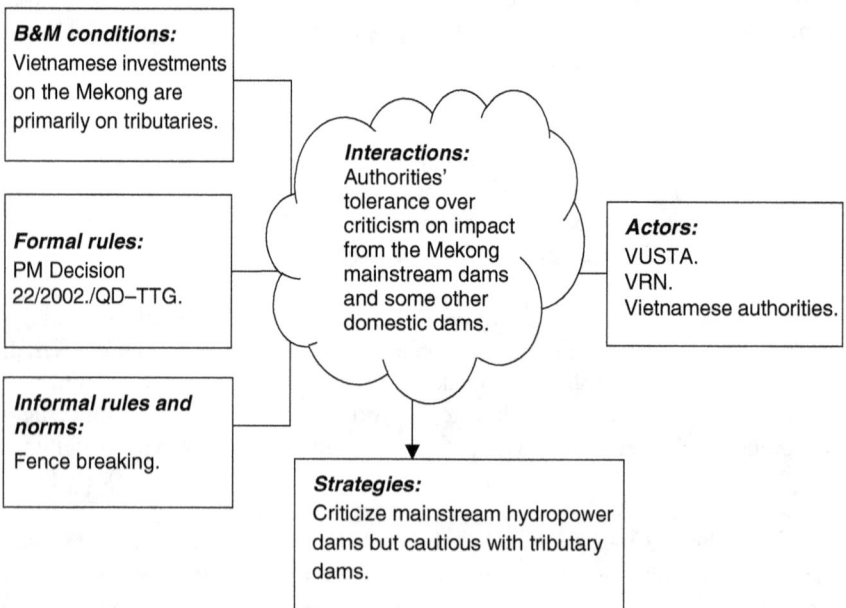

Figure 10.1 Influence of biophysical and material conditions, formal rules, informal rules and norms, and actors on VRN's strategy in raising hydropower dam issues with the authorities

This analysis of the Vietnamese situation is summarized in Figure 10.1. Formal rules such as the Prime Minister's Decision 22/2002/QD-TTg, which provides a mandate for VUSTA to critically review the government's policies, enabled a situation in which VUSTA could conduct reviews of the impacts of domestic hydropower dams. This situation, along with the Vietnamese fence-breaking culture, allowed a certain level of criticism over the authority's actions, and created a situation in which the Vietnamese government was tolerant towards a certain level of criticism on hydropower dams. However, the fact that Vietnam was investing in dams along the tributaries of the Mekong River did not allow criticism of these developments. The analysis illustrates how formal rules, informal rules and norms, actors, and biophysical and material conditions interacted and created the situation in which the VRN was able to criticize some hydropower dam issues, but not others.

In Cambodia, the RCC needed to be cautious when raising issues related to all hydropower dams with local and national authorities. As discussed in Chapter 9, the RCC also needed to be sensitive when raising the issue of hydropower with the Council of Ministers (C5 2012). At the local level the dominant influence over the RCC's activities was the informal pressure exerted by local authorities not to speak up against hydropower development, combined the fact that people were afraid of speaking up against the local authorities (C12 2012; C15 2012). It is interesting to note that this hesitancy and pressure persisted at the local level even after the Cambodian government officially announced its concerns over the Xayaburi dam (Kingdom of Cambodia 2011).

The informal pressure not to speak up against the authorities arises from various factors, as illustrated in Figure 10.2. As discussed in Chapter 6, the current political situation where people are afraid to speak up against authorities was created through manipulation by the current ruling party of the decentralized governance system, which is based on formal rules. The current ruling party also used informal rules and norms such as neo-patrimonialism. In addition, traditional norms based on Theravada Buddhism's teachings, which do not allow people to speak up against others who have gained higher social status, and the traditional fear of authorities enhanced by the historical legacy of Khmer Rouge, contributed to creating the situation in which some RCC and community members were afraid of speaking up in the face of authority.

The norm of not speaking up against the authorities contradicts the formal rule of freedom of expression for all Cambodians which is guaranteed in the Cambodian Constitution (Article 41 *Constitution of Cambodia* 1993). Cambodia has adopted a multi-party political system, and the right of citizens to establish with political parties is also guaranteed in the Constitution (Article 42 *Constitution of Cambodia* 1993). However, in reality, exercising this right in Cambodia is extremely difficult. The Cambodian People's Party (CPP), led by Prime Minister Hun Sen, dominates the political scene, and some scholars claim that Cambodia is moving towards an authoritarian state under his leadership (McCargo 2005; Heder 2005; Un 2011). The major opposition party leader Sam Rainsy is in exile, and in 2010 he was sentenced *in absentia* to

B&M conditions:
Cambodia's plans for
hydropower dams
on the mainstream
of the Mekong River.

Formal rules:
Policies and laws on
decentralization.
Cambodian
government's
intention to introduce
LANGO.

**Informal rules and
norms:**
Neo-patrimonialism.
Fear of authorities.
Legacies of spies.
Taboo in criticising
people who gained
higher social status
(Theravada Buddhism).

Interactions:
Informal pressure not to
speak up against authorities
and development.

NGOs need to maintain
positive relationship with
local authorities.

Actors:
CPP.
Local authorities.
RCC member
NGOs (mix of
advocacy and
development-
focused
organizations).

Strategies:
RCC's cautious approach in raising
the issue of hydropower dams.

Modification of planned activities –
some RCC/community members
did not take part in thumbprint
activities.

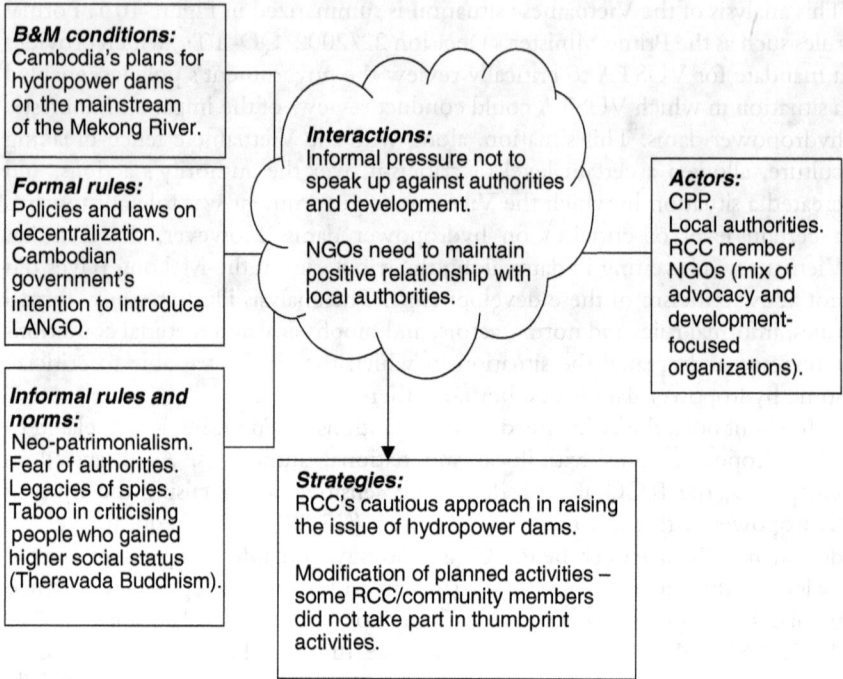

Figure 10.2 Influence of biophysical and material conditions, formal rules, informal rules and norms, and actors on the RCC's strategy in raising the issue of hydropower dams with the authorities

11 years in prison for a number of charges which appear to be politically motivated (Human Rights Watch 2010; Un 2011: 557; *BBC News Asia* 2013; United Nations Human Rights Council 2012: 14). As discussed in Chapter 6, this political situation is also resulting in fear of the authorities.

In addition, the plans of the Cambodian authorities to build hydropower dams on the Mekong River created a sensitive situation with local authorities. As discussed in Chapter 4, Cambodia has plans to build two hydropower dams on the mainstream of the Mekong River in addition to other dams on the tributaries of the Mekong. This development was of special concern for Stung Treng and Kratie provinces, where two hydropower dams are planned on the mainstream of the river. These are expected to cause the largest impacts on fish production as compared with other dams planned on the mainstream of the Mekong River (ICEM 2010: 103). The planned building of these dams created the situation where local authorities in these provinces were sensitive to any discussions by NGOs on the impacts of the hydropower dams.

Another key factor that created informal pressure by local authorities was associated with the profile of RCC's member NGOs, which was a mix of advocacy- and development-focused NGOs. While attempting to ensure that development projects do not impact negatively on the rights and livelihoods of the communities with which they work, many of the development-focused NGOs find it essential to maintain positive relationships with government authorities, particularly the local authorities (C15 2012; C16 2012). This situation is not unique to the RCC member NGOs. According to a survey conducted with 1409 CSOs (including 1010 local and 594 international organizations) in 2011, 39 per cent of the respondent CSOs identified local authorities as primary stakeholders (Cooperation Committee for Cambodia 2012a: 47). It is interesting to note that the same survey illustrated that CSOs considered relationships with local authorities more important than those with central government (Cooperation Committee for Cambodia 2012a: 48). One interviewee commented that this may be due to the geographic location of NGOs, as Phnom Penh-based NGOs tend to consider central government as an important stakeholder, whereas local authorities are more important stakeholders for province-based NGOs (C30 2012).

The survey by the CCC indicates that while many CSOs are registered in Phnom Penh, each respondent CSO operates on average in 3.7 provinces (Cooperation Committee for Cambodia 2012a: 44). Due to the importance of maintaining positive relationships with the local authorities, development-focused NGOs fear jeopardizing these relationships by engaging in advocacy work. These NGOs also wish to avoid being seen as those opposed to government development initiatives (C16 2012; C15 2012). One interviewee indicated that most NGOs are not critical of the government; the exception is advocacy NGOs, which are often supported by donors who expect them to be critical (C25 2012: 25). Advocacy NGOs are a minority in Cambodia, comprising approximately 7 per cent of all NGOs in the country (Bañez-Ockelford 2010: 16). This factor provides an explanation of the caution displayed by some of the RCC member NGOs when engaging in local-level advocacy work such as the collection of thumb prints (C15 2012; C16 2012; C12 2012; C13 2012).

The importance of maintaining positive relationships with the authorities is illustrated in the working arrangements between NGOs and government in Cambodia. Typically, advocacy-oriented NGOs and NGOs supporting human rights do not have official working arrangements beyond formal registration under the Ministry of the Interior (MoI) (C35 2012). However, most NGOs working on development-oriented projects typically have either a Memorandum of Understanding (MoU) or a counterpart from the government institutions with the most relevant mandates or areas of responsibilities for the NGOs' projects (C35 2012; C30 2012). These arrangements are meant to facilitate the work of NGOs, particularly with the government. However, in reality the most important factor associated with this collaboration relates to sharing financial and human resources associated with projects implemented by NGOs (C35 2012; C30 2012). Having a government counterpart typically requires NGOs to provide financial incentives to a government officer, who in return allocates some of his or her working time to the project (C35 2012; C30 2012). One of the interviewees commented that having government counterparts inside the organization restricts the NGO's ability to be critical of the government (C30 2012). On the other hand, the MoU arrangement typically allows two institutions (NGO and government institution) to allocate project funding to the institutions, rather than to specific individuals (C35 2012; C30 2012). While Cambodian NGOs are designed to conduct their activities independently from the government, in reality their activities are somewhat restricted through these formal arrangements.

Working for local NGOs has typically been considered to be a 'stepping stone' in one's career, particularly by young university graduates (C27 2012). Many Cambodian graduates start as volunteers at local and international NGOs, and gradually move on to obtain positions within NGOs. When they find a position at another organization with a better salary, they move on to the next step. Speaking against the government in Cambodia is associated with the risk of personal threat to many NGO staff members (C13 2012; C10 2012). Some RCC members faced personal threats through their anti-dam activities (C27 2012; R2 2012). In a recent case, an outspoken Cambodian activist, Chut Wutty, who was investigating illegal logging associated with the dam construction in Koh Kong province, was suspected of having been killed by the military (Amnesty International 2012; Open Development 2012). For NGO staff members who happen to work on advocacy projects as part of their career trajectory, engaging in advocacy that opposes the government may not be worth the risk to their personal security (C28 2012). Therefore many of them tend to take conservative positions towards the government rather than counter the authorities, one reason for 'fear' towards authority.

As discussed in Chapter 6, the Cambodian government's attempt to introduce the Law on Associations and Non-Governmental Organizations (LANGO) is considered as a way to restrict community movements and to prevent civil society from using its freedom of association, which is guaranteed in the Cambodian Constitution (Article 42 *Constitution of Cambodia* 1993; Amnesty International 2011; Cooperation Committee for Cambodia 2012b). According to the draft version of LANGO, if associations or NGOs do not register with the MoI or sign a memorandum they are not allowed to operate within Cambodia (Article 6 *LANGO (3rd Draft)* 2011). This may pose a challenge for community-based organizations and small-scale NGOs that may not have the capability and funds to arrange for registration. The introduction of LANGO may be politically motivated by the CPP to maintain its political dominance. In order to sustain its political life, it is important for the CPP to eliminate any movements that may threaten its power. A community movement could potentially be one such, as it could potentially threaten the CPP's power, particularly in rural areas where the CPP has its strongest political support base (C21 2012). Intimidation of community activists exists already, as one RCC member based in a rural area indicated. When community members travel to Phnom Penh to attend a seminar or dialogue related to hydropower dams, they are at times stopped by police *en route* to Phnom Penh, or confined to their guesthouse by police upon reaching the capital (C14 2012). Supporting communities and advocating for their rights could also impact on NGOs, as illustrated in the case where the MoI suspended the Cambodian NGO which was supporting poor families evicted by the government-led railway project (Peter 2012). These pressures through formal rules and informal practices contribute to the informal pressure for NGOs not to speak up against the government.

In summary, a combination of formal rules, informal rules and norms, actors, and physical and material conditions associated with the Mekong's hydropower dams contributed to creating some RCC members' hesitancy in raising certain issues (in this case, hydropower dams) in the eyes of local authorities.

NGO–community relationships

One factor that shaped the way both the Vietnamese and Cambodian coalitions worked with communities relates to their existing relationships with the stakeholders in the potentially affected areas. In Cambodia, the RCC member NGOs are based in various provincial towns along the Mekong River, and most have ongoing field projects with communities in the area. When the RCC members conducted awareness-raising events or provided information about the dam to the communities, these existing networks were used as a pathway of communication as well as a means of organizing events. The engagement of these development NGOs was relatively new. The original membership of the Se San Working Group (RCC's predecessor) was primarily composed of advocacy NGOs (such as 3SPN, NGO Forum, FACT). However, when the RCC expanded the geographic scope of its work to all dam issues throughout Cambodia, it also expanded its membership to NGOs based in different parts of Cambodia whose main goals were to improve livelihoods, education, and natural resources for rural Cambodians (NGO Forum on Cambodia 2011). Many of these NGOs joined the RCC as hydropower dams are one of the key development issues affecting the livelihoods of the communities to whom they provide services (C13 2012; C15 2012; C16 2012). The RCC members' close and ongoing relationship with community members was a key factor allowing the RCC to conduct a variety of community awareness-raising activities associated with the Xayaburi hydropower dam.

Another characteristic related to Cambodian NGOs is the fact that many of them do not have constituencies (C21 2012; Bañez-Ockelford 2010: 20). As discussed in Chapter 6, most Cambodian NGOs emerged as a result of the sudden availability of international funding and the need to implement donor agendas after the Paris Peace Agreement was signed in 1991 (Un 2006: 242; Bañez-Ockelford 2010: 4). They did not appear as a result of a gradual emergence of social capital, nor the gradual scaling-up of grassroots organizations (Un 2006: 242; Malena and Chhim 2009). According to one interviewee, even the NGO Forum did not allow community groups to be part of their network in the early days (C21 2012). Some of the RCC member NGOs emerged through community movements such as FACT and 3SPN, and later institutionalized themselves as NGOs registered under the MoI (Fisheries

Action Coalition Team 2012: 9; C6 2012; 3S Rivers Protection Network 2013). However, these constituency-based NGOs are a minority within the RCC network (NGO Forum on Cambodia 2011).

As a result, local communities are typically considered as beneficiaries of NGOs' work rather than as their constituencies. While most projects take a participatory approach to determining project activities, there are often certain objectives that NGOs need to achieve through commitment of deliverables to donors. A survey conducted in 2011 by CCC reflects this fact, as more CSOs identified donors as their primary stakeholders (65 per cent) than their community beneficiaries (48 per cent) (Cooperation Committee for Cambodia 2012a: 47). Donors tend to have a large influence on determining the strategic focus and directions of NGOs in Cambodia (Bañez-Ockelford 2010: 32). A traditional patron–client relationship is in some ways reflected in this current donor–NGO relationship in Cambodia (Bañez-Ockelford 2010: 32; C21 2012). This characteristic of Cambodian NGOs is also common to many of the RCC member NGOs, as many of the member NGOs indicated that they determine their activities associated with hydropower advocacy in consultation with their donors, rather than with community members (C9 2012b; C13 2012; C28 2012).

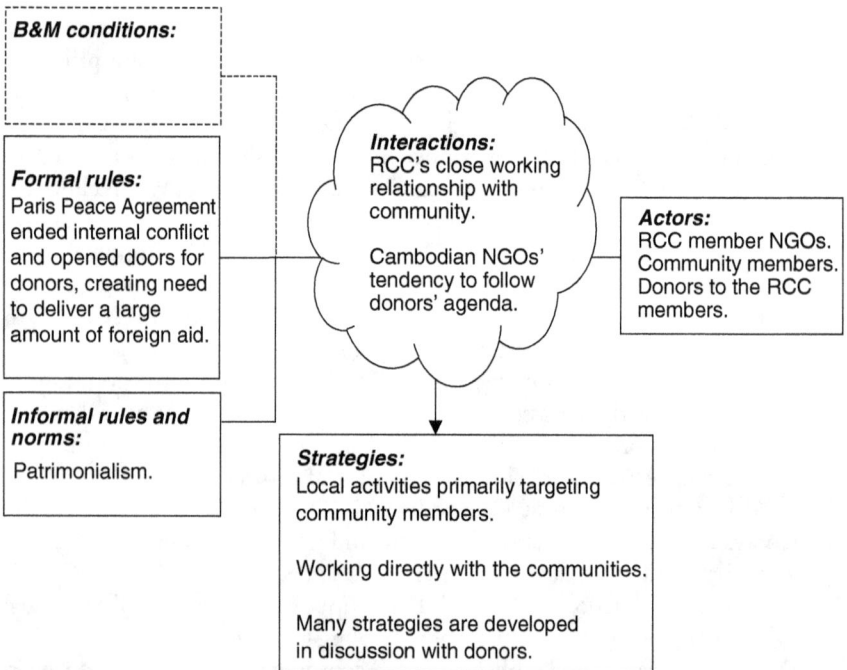

Figure 10.3 Influence of formal rules, informal rules and norms, and actors on RCC's strategy of targeting communities

Figure 10.3 illustrates how the RCC's strategies in Cambodia were influenced by interaction of actors and underlying influence of formal and informal rules and norms. Two key factors associated with the actors had direct influence on the RCC's strategies. First, the RCC members' characteristics in working closely with community members influenced its focus on working with community awareness-raising. The second factor is the RCC members' close relationship with their donors in shaping the activities. In addition, two factors that affected NGO–donor relationships historically in Cambodia may have affected the way the RCC developed its strategies with communities. One is the Paris Peace Agreement (1991): the formal rule which was signed to end the internal conflict in Cambodia created the situation where a large number of Cambodian NGOs mushroomed to implement donor agendas. The other is the patron–client relationship, which is based on the Cambodian traditional norm of patrimonialism that created the tendency for NGOs to follow donors' interests and agendas.

In contrast to the RCC, which had direct contacts with stakeholders in affected areas through the member NGOs' ongoing relationship with communities, the VRN used two different pathways for its work with stakeholders in the Mekong Delta. One was through utilizing personal networks of VRN members. The VRN members working at Can Tho University played a key role in organizing workshops with farmers within the Mekong Delta (V2 2012; V16 2012). They were said to have decades of collaboration and working relationships with the local farmers, which also worked to the advantage of the VRN's activity (V17 2012). One of the VRN members interviewed indicated that many of the provincial government staff graduated from Can Tho University, which allowed the VRN members to have a good working relationship with alumni working at local government offices (V17 2012).

Another pathway was the use of existing organizations that are closely related to the state. The workshop targeting local government officials was organized through the Southwest Steering Committee, a regional coordination body of provinces within the Mekong Delta (V15 2012; V2 2012). The Committee is chaired by the Vice Minister and located directly under a political bureau within the Vietnam Communist Party (V15 2012). Being a politically important coordination body, the Southwest Steering Committee was able to write a letter to the central government after the workshop, raising concerns about the Delta provinces (V16 2012; Southwest Steering Committee 2011). The collaboration with the Southwest Steering Committee was possible because a member of ForWet, an NGO based in Ho Chi Minh City, knew someone from the Committee (V15 2012; V16 2012). Here again, the use of an existing network and its communication channels to the central government through local organizations was seen as important.

Collaboration with farmers' unions was another way to invite farmers to the workshop (V16 2012). As discussed in Chapter 5, farmers' unions are one of the mass organizations under the umbrella of the Vietnam Fatherland Front (VFF)

(Norlund *et al.* 2006: 33). The unions have close link with the Communist Party of Vietnam (CPV) (Viet Nam Farmers' Union undated). This means that in approaching communities, the VRN worked with existing organizations closely linked with the state. In understanding the NGO–community relationship, it is important to consider this relationship in the overall context of state–civil society relationships and the positioning of NGOs within this context. As discussed in Chapter 5, in Vietnam mass organizations have represented civil society since the 1930s, and there were no other CSOs until Vietnamese NGOs emerged in the late 1980s (Norlund *et al.* 2006). When Vietnam opened its doors to a western market economy after the introduction of the *Doi Moi* policy, some donors engaged with mass organizations in lieu of local NGOs in the western sense, due to the lack of existing local NGOs (Thayer 2008: 8). Compared with the mass organizations, Vietnamese NGOs (VNGOs) are relative newcomers at the local level in Vietnam, and have not yet gained the same level of popularity and trust within the local population (Norlund *et al.* 2006: 111). One interviewee commented that, due to this general lack of trust, community members may be more cautious with NGOs as opposed to government agencies (V12 2012). The way the VRN conducted local-level activities in collaboration with farmers' unions reflects this reality in Vietnam.

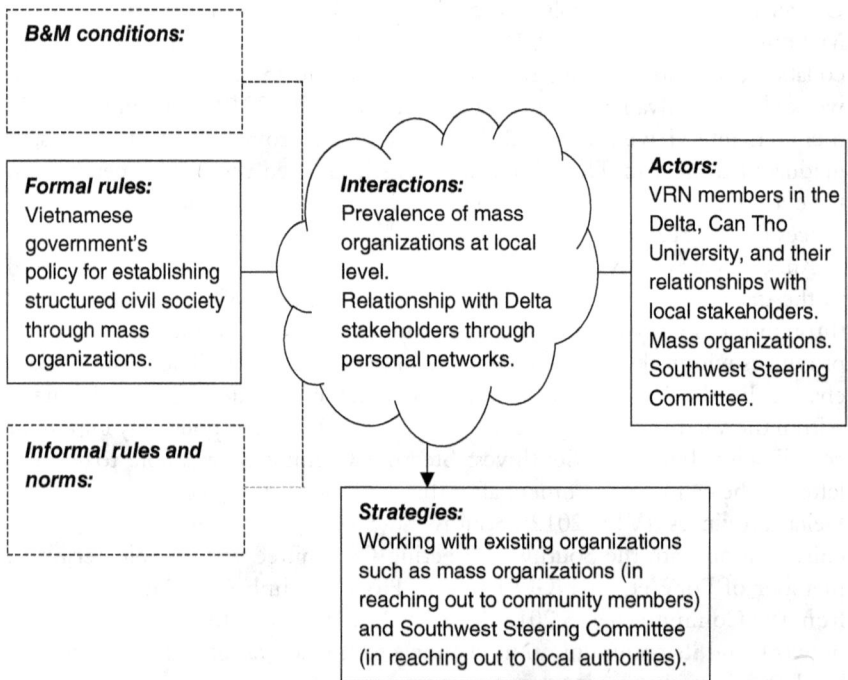

Figure 10.4 Influence of formal rules and actors on the VRN's strategy of targeting local stakeholders

Figure 10.4 illustrates the relationship between actors and the advocacy strategies of the VRN. The analysis shows that existing relationships among actors is one of the key factors that determined the strategies of the VRN. Another factor was the situation in which mass organizations were prevalent at the local level in Vietnam where the presence of VNGOs was limited. The prevalence of mass organization in Vietnam is a result of Vietnamese government policy in structuring civil society through government-led mass organizations from the 1950s (Sinh 2014: 43). This created the situation where VNGOs were newcomers in working with local stakeholders in Vietnam, and the VRN's collaboration with mass organizations was necessary in order to reach local community members.

Conclusion

This chapter analyses how rules and norms influenced NGO strategies targeting stakeholders within areas potentially affected by the Xayaburi hydropower dam. The analyses highlight three key points that support an understanding of the influence of rules and norms.

First, the analysis demonstrates that strategies were affected by actors' use of formal rules, informal rules, and norms. As clearly illustrated in the case of Cambodia, the formal political system of decentralization, which was defined by formal rules, has been manipulated through the use of informal rules and norms, namely neo-patrimonialism, in order to sustain the current government. This situation reflects competing relationships between formal rules and informal rules and norms, which aligns with the typologies of relationships suggested by Helmke and Levitsky (2004). As discussed in Chapter 2, Helmke and Levitsky suggest that where formal institutions are ineffective, the relationship is either substitutive or competing (Helmke and Levitsky 2004: 728). In Cambodia there is a competing relationship between formal rules, which guarantee freedom of expression including the right to vote, and informal rules and norms, which pressure people to behave in a certain way, namely to vote for the CPP and not to speak against the authorities. The existence of this competing relationship between formal rules and informal rules and norms also illustrates the fact that Cambodia is a legally plural landscape where competing forces of formal and informal rules exist (Gillespie 2011). As discussed by Helmke and Levitsky (2004), this relationship is determined by how actors adopt formal rules and informal rules and norms, as reflected in the Cambodian situation.

Second, physical and material conditions, namely the hydropower investment plans of respective national governments, were important factors that interacted with actors, formal rules, informal rules and norms in determining the sensitivity in the face of the authorities.

Finally, the networks and relationships among actors, particularly with local stakeholders, were another important factor that determined the way in which both NGO coalitions formulated their strategy. In the case of Cambodia, the relationship was established through member NGOs' ongoing activities and relationships with communities. In the case of Vietnam, the relationship with local stakeholders was developed partly through the existing formal structures of mass organizations, and partly through VRN members' relationships with stakeholders within the Mekong Delta.

References

3S Rivers Protection Network. 2011. *Living Rivers: News from the Sesan, Srepok and Sekong Rivers in Cambodia* 5: July–September. www.3spn.org/wp-content/uploads/2012/01/Living-Rivers-Newsletter-Edition-05_English.pdf. Accessed 26 October 2012.

—— 2013. *Background*. www.3spn.org/about-us/background. Accessed 23 November 2013.

Amnesty International. 2011. *Document-Cambodia: Proposed Law on Associations and Non-Governmental Organizations: A Watershed Moment?* www.refworld.org/docid/4ef862a72.html. Accessed 7 June 2015.

—— 2012. *Killing of Cambodian environment activist must be investigated*. www.amnesty.org/en/articles/news/2012/04/killing-cambodian-environment-activist-must-be-investigated. Accessed 7 June 2015.

Bañez-Ockelford, Jane. 2010. *Reflections, Challenges and Choices: 2010 Review of NGO Sector in Cambodia*. Phnom Penh: Cooperation Committee for Cambodia. www.ccc-cambodia.org/downloads/publications/2010_Review_NGO_Sector_Assessment.pdf. Accessed 7 June 2015.

BBC News Asia. 2013. Profile: Sam Rainsy. *BBC News Asia*, 26 July. www.bbc.co.uk/news/world-asia-23311394. Accessed 4 February 2014.

C3. 2012. Personal interview, 25 July 2012.

C5. 2012. Personal interview, 25 July 2012.

C6. 2012. Personal interview, 26 July 2012.

C8. 2012. Personal interview, 27 July 2012.

C9. 2012a. email communication to the author, 14 December 2012.

—— 2012b. Personal interview, 28 July 2012.

C10. 2012. Personal interview, 28 July 2012.

C12. 2012. Personal interview, 30 July 2012.

C13. 2012. Personal interview, 31 July 2012.

C14. 2012. Personal interview, 1 August 2012.

C15. 2012. Personal interview, 1 August 2012.

C16. 2012. Personal interview, 2 August 2012.

C17. 2012. Personal interview, 2 August 2012.

C18. 2012. Personal interview, 3 August 2012.

C20. 2012. Personal interview, 4 August 2012.

C21. 2012. Personal interview, 7 August 2012.

C23. 2012. Personal interview, 8 August 2012.

C25. 2012. Personal interview, 8 August 2012.

C27. 2012. Personal interview, 9 August 2012.

C28. 2012. Personal interview, 10 August 2012.

C29. 2012. Personal interview, 10 August 2012.

C30. 2012. Personal interview, 12 August 2012.

C35. 2012. Informal conversations with the author, September–October 2012.

Cambodian Rural Development Team. undated. *Projects.* www.crdt.org.kh/projects. Accessed 29 October 2012.

The Constitution of the Kingdom of Cambodia. 1993. Cambodia.

Cooperation Committee for Cambodia. 2012a. *CSO Contributions to the Development of Cambodia 2011.* Phnom Penh: Cooperation Committee for Cambodia. www.ccc-cambodia.org/downloads/publications/CSO_Contributions_2011.pdf. Accessed 7 June 2015.

—— 2012b. Law on Association and Non-Governmental Organizations (LANGO). www. ccc-cambodia.org/lango.html. Accessed 16 June 2012; no longer available online.

Fisheries Action Coalition Team. 2012. *Strategic Plan 2012–2014.* www.fact.org.kh/index. php?option=com_content&view=article&id=196%3Afact-strategic-plan-2012-2014&catid=58%3Apublication-a-reports&lang=km. Accessed 27 October 2012; no longer available online.

Gillespie, Josephine. 2011. Legal pluralism and world heritage management at Angkor, Cambodia. *Asia Pacific Journal of Environmental Law* 14 (1/2): 1–19.

Grimsditch, Mark. 2012. *3S Rivers Under Threat.* Berkeley, CA: International Rivers. www. internationalrivers.org/files/attached-files/3s_rivers_english.pdf. Accessed 7 April 2013.

Heder, Steve. 2005. Hun Sen's consolidation: death or beginning of reform? *Southeast Asian Affairs*: 113–130.

Helmke, G., and S. Levitsky. 2004. Informal institutions and comparative politics: a research agenda. *Perspectives on Politics* 2 (4): 725–740.

Human Rights Watch. 2010. *Cambodia: Opposition Leader Sam Rainsy's Trial a Farce.* www. hrw.org/news/2010/01/28/cambodia-opposition-leader-sam-rainsy-s-trial-farce. Accessed 27 May 2013.

ICEM. 2010. *MRC Strategic Environmental Assessment (SEA) of the Hydropower on the Mekong mainstream.* Hanoi: Mekong River Commission. www.mrcmekong.org/assets/ Publications/Consultations/SEA-Hydropower/SEA-Main-Final-Report.pdf. Accessed 3 May 2014.

Kingdom of Cambodia. 2011. *Mekong River Commission Procedures for Notification, Prior Consultation and Agreement: Form/Format for Reply to Prior Consultation.* Cambodia National Mekong Committee (CNMC). www.mrcmekong.org/assets/ Consultations/2010-Xayaburi/Cambodia-Reply-Form.pdf Accessed 25 November 2011.

Law on Associations and Non-Governmental Organizations (3rd Draft). 2011. Cambodia.

Lipes, Joshua. 2012. Hundreds Protest Lao Dam Project. *Radio Free Asia,* 29 June. www.rfa. org/english/news/cambodia/dam-06292012165509.html?searchterm=Xayaburi. Accessed 26 October 2012.

Malena, Carmen, and Kristina Chhim. 2009. *Linking Citizen and the State: An Assessment of Civil Society Contributions to Good Governance in Cambodia.* Phnom Penh: World Bank. http://siteresources.worldbank.org/EXTSOCIALDEVELOPMENT/Resour ces/244362-1265299949041/6766328-1307475897842/CCSA_Eng_Final.pdf. Accessed 4 October 2012.

McCargo, Duncan. 2005. Cambodia: getting away with authoritarianism? *Journal of Democracy* 16 (4).

NGO Forum on Cambodia. 2011. List of Rivers Coalition in Cambodia (RCC). (Document obtained from an interviewee.)

—— 2012a. Case study: RCC contributes to work with the RGC to call for cancellation of Xayaburi Dam. (Document obtained from an interviewee.)

—— 2012b. *Subject: Rivers Coalition in Cambodia Joint Statement: Call Thailand to Cancel the Power Purchase Agreement and Laos to stop construction and for both countries to respect the 1995 Mekong Agreement.* A letter addressed to H.E. Yingluck Shinawatra, Prime Minister of Thailand and H.E. Thongsing Thammavong, Prime Minister of Lao PDR, 27 July.

Ni, Duong Van. 2012. *Workshop Report: The impacts of hydropower dams on livelihoods in the Mekong Delta*, Vietnam Rivers Network Workshop, Can Tho, 26–27 November 2011.

Norlund, Irene, Dang Ngoc Dinh, Bach Tan Sinh, Dang Ngoc Quang, Do Bich Diem, Nguyen Manh Cuong, Tang The Cuong, and Vu Chi Mai. 2006. *The Emerging Civil Society: An Initial Assessment of Civil Society in Vietnam.* Hanoi: Vietnam Institute of Development Studies (VIDS), UNDP Vietnam, SNV Vietnam. www.civicus.org/new/media/CSI_Vietnam_report%20.pdf. Accessed 7 December 2011.

Open Development. 2012. *Chut Wutty Remembered in Koh Kong Memorial.* www.opendevelopmentcambodia.net/tag/natural-resource-protection-group. Accessed 28 October 2012.

Peter, Zsombor. 2012. New year brings return of suspended NGO. *The Cambodia Daily* 2 January.

Phuc, Dan Tiep. 2011. Hoa Binh hydropower: impacts on resettled and downstream communities. In *Water Resources and Sustainable Development: Perspectives from Laos and Vietnam*, Hanoi, 1–2 December 2011.

R2. 2012. Personal interview, 30 June 2012.

R8. 2012. Personal interview, 14 August 2012.

Sidel, Mark. 1997. The emergence of a voluntary sector and philanthropy in Vietnam: functions, legal regulation and prospects for the future. *Voluntas: International Journal of Voluntary and Nonprofit Organizations* 8 (3): 283–302.

Sinh, Bach Tan. 2014. Identifying civil society in Vietnam. In *Southeast Asia and the Civil Society Gaze: Scoping a Contested Concept in Cambodia and Vietnam*, edited by G. Waibel, J. Ehlert and H. N. Feuer. London: Routledge.

Southwest Steering Committee. 2011. *A Letter to: Office of the Government; Ministry of Foreign Affairs; Ministry of Natural Resources and Environment; Ministry of Agriculture and Rural Development; Vietnam National Mekong Committee; and Vietnam Union of Science and Technology Associations*, 5 December.

Sovuthy, Khy. 2012. Police detain man seeking justice for Sonando. *The Cambodia Daily* 28–29 July.

Thayer, C. A. 2008. One party rule and the challenge of civil society in Vietnam. In *Remaking the Vietnamese State: Implications for Vietnam and the Region*, Vietnam Workshop, Hong Kong, 21–22 August 2008.

Un, Kheang. 2006. State, society and democratic consolidation: the case of Cambodia. *Pacific Affairs* 79 (2): 225–245.

—— 2011. Cambodia: moving away from democracy? *International Political Science Review/ Revue internationale de science politique* 32 (5): 546–562.

United Nations Human Rights Council. 2012. *Report of the Special Rapporteur on the situation of human rights in Cambodia, Surya P. Subedi.* http://cambodia.ohchr.org/WebDOCs/DocReports/3-SG-RA-Reports/A-HRC-21-63_en.pdf. Accessed 26 May 2013.

V2. 2012. Personal interview, 10 July 2012.

V10. 2012. Personal interview, 14 July 2012.

V12. 2012. Personal interview, 16 July 2012.

V15. 2012. Personal interview, 18 July 2012.

V16. 2011. Personal interview, 23 November 2011.

—— 2012. Personal interview, 19 July 2012.

V17. 2012. Personal interview, 19 July 2012.

V20. 2011. Personal Interview, 1 December 2011.

V24. 2013. Informal conversations with the author, 16–21 June 2013.

Viet Nam Farmers' Union. undated. *Organization and operation of Vietnam Farmers' Union (VNFU)*. http://vnfu.vn/index.php/introduction/252-organization-and-operation-of-vietnam-farmers%E2%80%99-union-vnfu.html. Accessed 24 November 2013; no longer available online.

Vietnam Rivers Network. 2010. Giấy mời tham dự "Hội thảo chia sẻ kinh nghiệm quy hoạch phát triển ngành điện vùng Mê Kông- Việt Nam" (*An invitation to attend a workshop to share experiences of power development*) (Document obtained from an interviewee.)

VUSTA. 2006. *Much pressure on the process of people displacement, resettlement in Son La hydropower project*. www.vusta.vn/en/news/Vusta-Head-quarter/Much-pressure-on-the-process-of-people-displacement-resettlement-in-Son-La-hydropower-project-7310.html. Accessed 7 June 2015.

—— 2007. *Research on socio-economic environment in resettlement area of the Thac Ba hydropower plant after 32 years' construction of the dam*. www.vusta.vn/english3/news/?16030/Research-on-socio---economic-environment-in-resettlement-area-of-the-Thac-Ba-hydropower-plant-after-32-years'-construction-of-the-dam.htm. Accessed 26 June 2013; no longer available online.

VUSTA, WARECOD, VRN, VNMP, CIWAREM, and CEWAREC. 2011. *Press release: Open Dialogue 'Xayaburi dam and hydropower cascade development in the mainstream of the Lower Mekong river basin: GAINS-LOSSES to Vietnam and downstream countries*.

Wyatt, Andrew B., and Ian G. Baird. 2007. Transboundary impact assessment in the Sesan River Basin: the case of the Yali Falls Dam. *International Journal of Water Resources Development* 23 (3): 427–442.

11 Strategies targeting the general public

Introduction

Raising the general public's awareness of issues is important in advocacy work as public opinion can potentially influence decision-making. In a democratic regime, in principle, citizens have the right to express their views about politics and political decision-making through their votes (Roskin *et al.* 2008; Dalton 2006: 201). Public opinion can influence the voting preferences of constituencies, and this is an important concern for the politicians whose political lives are dependent on re-election (Dalton 2006: 206).

Public opinion can also create norms within societies which create common understandings of the importance of issues, and can facilitate the emergence, adoption, and implementation of new policy and law (Brown *et al.* 2000: 20). NGOs often play critical roles as 'norm entrepreneurs' in the process of promoting norm development (Koh 1998). Norm entrepreneurs are the agents who actively build notions about what is appropriate or desirable behaviour in their communities (Finnemore and Sikkink 1998: 896). The way norms are framed and interpreted affects their acceptance by society at large (Finnemore and Sikkink 1998: 897). Framing is also an important concept for the media as it creates a storyline, providing meanings to events being communicated (Scheufele 1999: 106). The way stories are framed in the media can affect citizens' judgements of the issue being reported, thus affecting public opinion (Simon and Jerit 2007: 258).

Advocacy organizations adopt different strategies to raise public awareness on issues they advocate for. When they have direct access to the target audience, advocacy organizations can conduct awareness-raising events such as training, community meetings, and distribution of posters and leaflets (Gordon 2002: 43). When advocacy organizations do not have direct access to the target audience, they can use media such as radio, newspapers, television, and the internet to disseminate information (Gordon 2002: 43). The key strategy adopted by the case-study coalitions for raising awareness among the general public (as opposed to stakeholders) over the Xayaburi dam was the use of media. Therefore this chapter examines the media strategy in detail, and analyses whether, and how, rules and norms shaped these strategies. The chapter first discusses the strategies adopted by both the Vietnam Rivers Network (VRN) and the Rivers Coalition in Cambodia (RCC); this discussion is followed by an analysis of how rules and norms influenced these strategies.

Strategies adopted by the VRN

The VRN used media as a way to promote issues of the Xayaburi dam among the general Vietnamese public (V2 2012; V11 2012; V1 2012). Some interviewees commented that the Vietnamese media play an important role in shaping the opinions of Vietnamese citizens (V11 2012; V9 2012). The interviewees also commented that Vietnamese citizens tend to listen to and trust most information issued by the media (V9 2012; V11 2012), despite the fact that what goes into the media is strictly restricted and monitored by the government (Wagstaff 2010; McKinley 2011: 94), as discussed later in this section. According to McKinley, Vietnamese newspapers have served as a tool for propaganda by the communist regime, and people tend to believe unquestioningly what they read in newspapers (McKinley 2011: 95). This practice made newspapers an important means of guiding public opinion that influenced the government's media policy (McKinley 2011: 95). One interviewee commented that Vietnamese leaders listen to public opinion as they need to be seen as responsible and want to be re-elected (V11 2012). According to a study that examined the relationship between press freedom and confidence in governments globally, Vietnam, which has a low level of press freedom, had a high level of confidence in government (Norris and Inglehart 2010: 207). This indicates that formulating one's story in the Vietnamese media should be an effective way to raise public interest and awareness. Therefore targeting the media was a strategic entry point for the VRN in its efforts to raise public awareness of the Xayaburi dam. This potentially affects the opinions of the decision-makers who are concerned about public opinion (V11 2012).

The VRN members provided information on the Xayaburi dam directly to journalists through personal contacts as well as through media releases (V2 2012; V16 2012; V1 2012). The VRN was one of the most important sources of information for journalists reporting on the Xayaburi dam, particularly as Vietnamese journalists have limited capacity and budget to conduct their own research and investigations on international issues (V9 2012; V19 2012). The VRN members also conducted workshops targeted at journalists. In addition to the two workshops conducted in Ho Chi Minh City, targeting government officials, scientists, and journalists (discussed in Chapter 9), the VRN also conducted field trips for journalists to help them to understand issues related to the Mekong Delta (V10 2012). Two field trips were organized during the dry and wet seasons (V10 2012).

The Centre for Water Resources Conservation and Development (WARECOD), the main coordinator of the VRN until 2012, also produced several films that highlighted issues relating to the Mekong Delta, which are available on YouTube (V10 2012; Van 2011). In addition, the VRN Mekong task force members also wrote articles for the newspapers (V16 2011).

The VRN's efforts to engage journalists faced barriers at times as a result of political interference with the media. In the early stage of the Xayaburi Procedures for Notification, Prior Consultation and Agreement (PNPCA) process, the Vietnamese government allowed the media to report freely about the Xayaburi dam (V9 2012; V19 2012). One interviewee commented that the government even encouraged the media to discuss hydropower construction on the Mekong (V9 2012). Shortly after the MRC Joint Committee meeting in April 2011, however, when the four member states could not reach consensus on the Xayaburi dam, journalists were told by the government not to write about the dam (V2 2012; V9 2012). This media embargo influenced the strategy of the VRN, which used the media as one means of disseminating information on the Xayaburi dam (V2 2012). When there was a media embargo on the Xayaburi dam, the media were not able to publish information relating to the dam, even when VRN members conducted related workshops or events and invited journalists, or submitted articles on the Xayaburi dam to the media for publication (V2 2012; V16 2012; V19 2012; V10 2012; V1 2012).

Strategies adopted by the RCC

A few Phnom Penh-based RCC members indicated that they used the media as an advocacy tactic to address the general public (C3 2012; C5 2012). The RCC's activities in raising public awareness through the use of media were conducted in a way that built on the existing programmes of its members. The Culture and Environment Preservation Association (CEPA), one of the founding member organizations of the RCC, conducted a regular radio programme highlighting the Mekong and indigenous issues (C3 2011). During the PNPCA process of the Xayaburi dam, the hydropower issue was discussed as part of this radio programme (C3 2011). The programme was broadcast through Women's Media FM 102, one of the few independent radio stations operating in Cambodia that is run by a Cambodian NGO (C35 2012; LICADHO 2008; Women's Media Centre of Cambodia 2012). The programme was broadcast six times a year, and affected communities from large-scale development projects were invited along with guest speakers from NGOs and the government (C3 2011). In the programme, communities raised issues of concern, followed by comments from NGOs and the government on their respective roles. Finally, independent analysts, mostly academics from universities, made comments and recommendations to different stakeholders (C3 2011).

The RCC also conducted other media-related activities such as issuing press releases (C5 2012). Some RCC members also spoke about the Xayaburi dam on Cambodian TV programmes (Ath and Marks 2013). While the details of press releases and media coverage were not available from the RCC, one RCC member commented that 'we use the media to capture the issue, like climate change or hydropower issue, everything like that' (C5 2012). In addition to using the media, the RCC also conducted public forums as a way of raising public awareness, including conducting a youth forum on Xayaburi, and a national conference on climate change, agriculture, and energy (C3 2012; C5 2012; NGO Forum on Cambodia 2011).

Comparative analysis

Media censorship

Analysis of the Vietnamese case reveals the existence of censorship of the media by the authorities, while this was not specifically expressed by the interviewees in Cambodia. This was despite the fact that formal rules guarantee the freedom of expression in both countries (*Constitution of Vietnam* 2001; *Constitution of Cambodia* 1993). However, in reality, the media are censored in both contexts.

Why was there a media embargo on the Xayaburi dam in Vietnam? Many interviewees indicated that this was because the Vietnamese government did not want to raise any tension with its counterpart in Laos (V19 2012; V9 2012; V2 2012). Politically, Vietnam and Laos have had a 'special relationship' since the establishment of the communist regime in Laos (Thayer 1982: 245). The close relationship and collaboration between these two countries politically, economically, and militarily has been formalized as the Lao–Vietnamese Treaty of Friendship and Co-operation (1977) (Thayer 1982: 253). However, in recent years there has been an increase in China's influence over Laos (Fujimura 2010), an influence that competes with that of Vietnam, thus threatening this relationship (V9 2012). The Vietnamese government was particularly interested in maintaining its collaboration with Laos as there is increasing competition with China over the territorial disputes between China and Vietnam within the South China Sea (Thayer 2012; *BBC News Asia* 2013; V9 2012; V19 2012). Interviewees commented that due to this sensitivity Vietnam was very careful not to offend Laos publicly, thus resulting in the media embargo on the Xayaburi dam (V19 2012; V16 2012).

This political interference on media freedom in Vietnam illustrates the contradictory relationship among formal rules and informal rules and norms. The Vietnamese Constitution guarantees freedom of expression by Vietnamese citizens (Article 69 *Constitution of Vietnam* 2001; Rieu-Clarke and Allan 2008: 26). At the same time, the Constitution designates the state as responsible for promoting mass communication, and bans mass communications that are 'detrimental to national interests, and destructive to the personality, morals and fine lifeway of the Vietnamese' (Article 33 *Constitution of Vietnam* 2001). The Law on Media assures freedom of media and speech (Article 2), but at the same time bans the media from reporting on issues that would incite people to rebel against the state and to disclose state secrets (Article 10) (*Law on Media* 1999).

In reality, the media struggle to exercise their freedom in Vietnam. Existing literature indicates that interventions on the content of publications and radio and television broadcasts by the Ministry of Culture and Information and the Communist Party's Department for Culture and Ideology are common practices in Vietnam (Kerkvliet 2001: 252; Matsumoto 1999: 126; Hayton 2010). This was echoed by the Vietnamese journalists interviewed by the author, who commented that the Ministry of Information and Communication and the Communist Party's Central Committee Commission on Popularization and Education conduct regular meetings with newspaper editors, informing them of what cannot be written in the Vietnamese press (V9 2012; V19 2012).

Editors-in-chief are held responsible if a newspaper violates the order or crosses the party line that is deemed acceptable by the Vietnamese Communist Party (Matsumoto 1999: 126; V9 2012; V19 2012). They face the threat of legal prosecution, as the Penal Code includes vaguely worded terminology that prohibits the freedom of speech and journalism if it significantly infringes on the interests of the state (Article 258 *Penal Code, Vietnam* 1999; Freedom House 2011). The Law on Media specifically indicates that heads of media organizations and journalists would be punished for breaching Article 10 of the Law on Media, which details matters not permitted to be reported in the media (Article 28 *Law on Media* 1999). As an example, journalists who tried to expose a corruption scandal in 2006 by a project management unit in the Ministry of Transport were arrested and charged with abuse of power (Thayer 2008: 18).

The analysis of the media activities conducted by the VRN illustrates the way formal rules, informal rules and norms, and actors, mostly at the Vietnamese

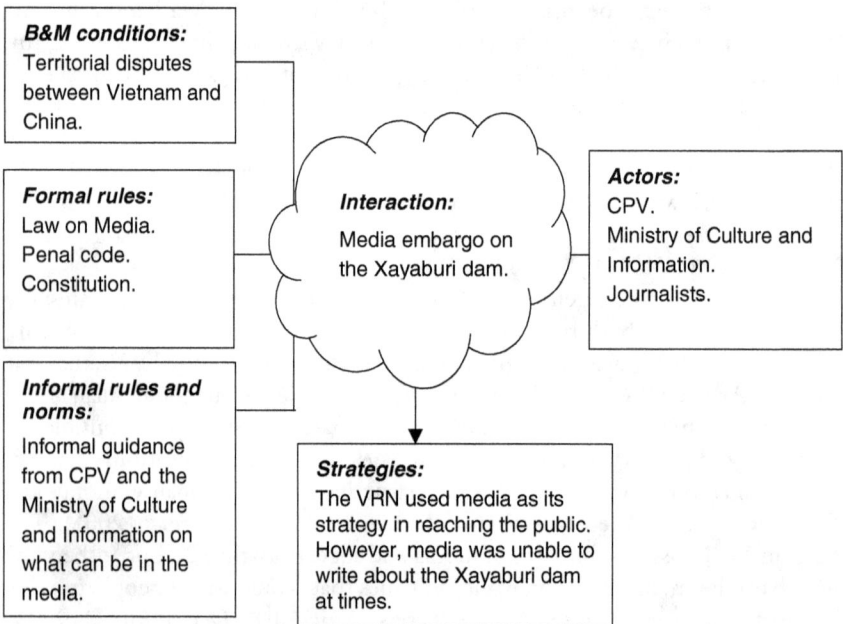

Figure 11.1 Influence of biophysical and material conditions, formal rules, informal rules and norms, and actors on VRN's media strategy

national level, interact with each other and influence the VRN's activities. This interaction is illustrated in Figure 11.1. Despite the existence of formal rules such as the Constitution and the Law on Media, which guarantee freedom of expression, informal rules restricting what can and cannot be publicized prevail. The informal rules are imposed on journalists by formal organizations such as the Communist Party of Vietnam (CPV) and the Ministry of Information and Communication. The Penal Code, a formal rule that restricts freedom of speech and press which infringe on the interest of the state (Article 258 *Penal Code, Vietnam* 1999), adds to this contradiction. These formal and informal rules exercised by actors put pressure on journalists, particularly on editors-in-chief, to highlight or restrict certain issues in the press. This informal pressure in the case of the Xayaburi dam was considered to be in place primarily due to territorial disputes between China and Vietnam, representing influence of biophysical and material conditions.

In Cambodia, the RCC members interviewed did not particularly mention constraints in their activities associated with the use of media, and no clear link associating rules and norms and the RCC's media strategy were observed through this case study. However, literature and interviews revealed the general constraints that journalists face in Cambodia (C25 2012; CCIM 2013; Samnang 2002). These include 'fear' of speaking up against authorities, as discussed in Chapter 7. The Cambodian Constitution guarantees the freedom of the press (Article 41 *Constitution of Cambodia* 1993). However, the Cambodian press is not in reality free (Freedom House 2012b). Television and radio are the main sources of information for many Cambodians as illiteracy is high, particularly in the rural areas (C25 2012; CCIM 2013). As a way to control information, all television stations and most radio stations are either owned or controlled by the ruling party (Freedom House 2012b; Un 2011: 552; CCIM 2013). The few radio stations operating independently from the ruling party frequently face harassment and intimidation from the government (Un 2011: 552; C25 2012; CCIM 2013: 10). Mam Sonando, the owner of Beehive Radio which is one of the independent radio stations, repeatedly faced charges by the government (LICADHO 2008: 38). At the time of author's field work (July 2012) he was arrested after broadcasting on a complaint lodged at the International Criminal Court accusing the Cambodian government of committing crimes against humanity in reference to forced evictions (Seiff and Sovuthy 2012: 1). He was later sentenced to 20 years in jail (BBC News Asia 2012). Physical attacks on journalists have decreased over time; however, threats aimed at journalists have shifted to the courtroom (C25 2012; Freedom House 2012b). Journalists who are critical of governments are often charged with defamation and disinformation (LICADHO 2010: 25; C25 2012).

While the Law on the Press provides some safeguards for journalists, such as a ban on censorship and confidentiality of sources, it also has loopholes which allow charges against and convictions of journalists (Un 2011: 552; Article 3 *Law on the Press* 1995). Article 12 of the Law on the Press prohibits publishing any information that 'may affect national security and political stability' without giving a clear interpretation of what this means (Article 12 *Law on the Press* 1995). The Penal Code (2009), which became effective in 2010, criminalizes defamation, and is often used to penalize journalists for criticizing high-level government officials (Articles 305–310 *Penal Code, Cambodia* 2009; Un 2011: 552; Freedom House 2012b; Article 10 *Law on the Press* 1995). These laws, in addition to the corrupt judiciary in Cambodia (Subedi 2011: 254), create threats for journalists, limiting the freedom of the press (Freedom House 2012a). These cases illustrate the interactions between formal and informal rules that result in creating 'fear' of the authorities.

The way the media are censored differs in each context. In Vietnam, there is clear direction from the CPV on which issues can or cannot be discussed in the media (V9 2012; V19 2012). In Cambodia, the line is not clear, and journalists use their own judgement and risk assessment regarding what would be acceptable to the ruling party (C25 2012). Unlike the Vietnamese government, which provides regular guidance on what should and should not be reported in the media, the Cambodian government does not give journalists similar guidance (LICADHO 2008: 46). However, due to fear of physical threats and legal prosecution, many journalists impose self-censorship (CCIM 2013: 8; LICADHO 2008).

This difference in the way censorship takes place is associated with the political system of each country. Vietnam is a socialist country, which follows Marxist–Leninist thought in its nation's foundation (Preamble *Constitution of Vietnam* 2001). Soviet media theory, which derives from Marx and Engels, claims that in the socialist context the working class holds power in society, and in order to keep power the means of 'mental production' needs to be controlled by the representatives of the working class, which is seen as the communist party (McQuail 1987: 118). This theory theoretically justifies censorship and punishment if the media are against the government (McQuail 1987: 119). The Constitution also clearly defines the state's role in promoting the mass media and, at the same time, controlling mass information which is detrimental to national interests (Article 33 *Constitution of Vietnam* 2001). On one hand, informal guidance from the state authorities on restriction of the media contradicts the freedom of speech guaranteed by the Constitutions of both Cambodia and Vietnam. On the other hand, imposing restrictions on the press and other media is justified through a theoretical basis for media restriction in the socialist context, and associated laws support this restriction (*Constitution of Vietnam* 2001; McQuail 1987).

In contrast, Cambodia's political system is a democracy. The democratic-participant theory of the media lies in claims that the media exist in order to satisfy the needs, interests, and aspirations of active receivers of information, and the organization and content of the media should not be subject to political or state control (McQuail 1987: 122–123). While infringing on national security or other people's fame is used in Cambodia as a basis for pressurizing journalists, the Constitution guarantees the freedom of the press and does not provide a clear responsibility for the state to restrict mass media that infringe on national security (*Constitution of Cambodia* 1993). Compared with the Vietnamese Constitution, which includes the state's responsibility in promoting and controlling the media, the Cambodian Constitution is written in a way that emphasizes the importance of protecting this right for its citizens:

> Khmer citizens shall have freedom of expression, press, publication and assembly. No one shall exercise this right to infringe upon the rights of others, to effect the good traditions of the society, to violate public law and order and national security.
>
> (Article 41 *Constitution of Cambodia* 1993)

Access to media

The type of media used by each case coalition reflects differences in accessibility of the media in each country. As ways of communicating with the public, the RCC engaged with radio and TV, whereas the VRN worked with newspapers, TV stations, and radio (V1 2011; V2 2012; C3 2011; C5 2012; Ath and Marks 2013). The types of media used reflect public accessibility in each country. Newspaper circulations in Cambodia are limited mostly to Phnom Penh, reaching only 2 per cent of the population (CCIM 2013: 8). Radio and TV are the main sources of information for the wider population of Cambodia, and thus are open to government censorship (C25 2012; CCIM 2013). In Vietnam, TV is the most popular medium, with approximately 90 percent of households owning a TV (Wagstaff 2010: 108). However, newspapers are also widespread throughout Vietnam (V19 2012; V11 2012). As of 2010 there were 178 newspapers circulating in Vietnam; this number includes 76 national papers and 102 local papers (McKinley 2011: 95). This difference in newspaper circulation between the two countries may also be the result of differences in literacy rates. The adult literacy rate in Cambodia was 77.6 per cent in 2008, whereas in Vietnam it was 90 per cent as of 2009 (United Nations Vietnam 2010; UNESCO Office in Phnom Penh undated). While social media are emerging media for mass communication, none of the interviewees particularly mentioned the use of new types of social media (including Facebook, Twitter, websites and blogs), except for the VRN, which uploaded to YouTube a video on changes in flood patterns in the Mekong Delta (V10 2012; Van 2011).

Conclusion

This chapter examines the NGO coalition's advocacy strategies in targeting the general public, particularly through the use of the media, and analyses how rules and norms influenced these strategies. The main issue observed in this analysis is the fact that in both Cambodia and Vietnam, freedom of expression and freedom of the press are at times suppressed by the authorities through both formal and informal rules. This situation indicates that formal and informal rules interacted in a way that created competing relationships. At times, one formal rule was used to counteract another, which also created competing relationships between formal rules. This interaction took place as a result of intervention by actors who are in a position to apply these rules.

The influence of these interactions on advocacy strategies was not observed in the case of the RCC. For the VRN, these influences were considered to be caused by the international relationships of China, Vietnam, and Laos. This situation indicates that actors' relationships count as another important factor that affected the way actors applied formal rules, informal rules, and norms.

References

1992 Constitution of the Socialist Republic of Vietnam: As Amended 25 December 2001 (transl. Allens Arthur Robinson). 2001. Vietnam.

Ath, Chhith Sam, and Danny Marks. 2013. *The NGO Forum's Campaign Against Xayaburi.* www.stimson.org/summaries/ngo-forums-campaign-against-xayaburi. Accessed 8 August 2013.

BBC News Asia. 2012. Cambodia jails journalist Mam Sonando over 'plot'. 1 October. www.bbc.co.uk/news/world-asia-19783123. Accessed 22 October 2012.

—— 2013. Q&A: South China Sea dispute. 22 January. www.bbc.co.uk/news/world-asia-pacific-13748349. Accessed 6 February 2013.

Brown, L.D., S. Khagram, M. H. Moore, and P. Frumkin. 2000. *Globalization, NGOs and Multi-sectoral Relations.* Working Paper No. 1. Cambridge, MA: Hauser Center for Nonprofit Organizations and Kennedy School of Government, Harvard University.

C3. 2011. Personal interview, 21 November 2011.

—— 2012. Personal interview, 25 July 2012.

C5. 2012. Personal interview, 25 July 2012.

C25. 2012. Personal interview, 8 August 2012.

C35. 2012. Informal conversations with the author, September–October 2012.

CCIM. 2013. *Challenges for Independent Media Development in Cambodia.* Phnom Penh: Cambodian Center for Independent Media. www.ccimcambodia.org/report/CCIM_report_indepedent_media_promotion.pdf. Accessed 11 August 2013.

The Constitution of the Kingdom of Cambodia. 1993. Cambodia.

Dalton, Russel J. 2006. *Citizen Politics: Public Opinion and Political Parties in Advanced Industrial Democracies,* 4th edn. Washington, DC: CQ Press.

Finnemore, Martha, and Kathryn Sikkink. 1998. International norm dynamics and political change. *International Organization* 52: 887–917.

Freedom House. 2011. *Vietnam: Freedom of Press 2011.* www.freedomhouse.org/report/freedom-press/2011/vietnam. Accessed 26 November 2012.

—— 2012a. *Freedom in the World: Cambodia.* www.freedomhouse.org/report/freedom-world/2012/cambodia. Accessed 10 October 2012.

—— 2012b. *Freedom of the Press 2012 – Cambodia.* www.unhcr.org/refworld/country,,,,KHM,,507bcae6c,0.html. Accessed 22 October 2012.

Fujimura, Kazuhiro. 2010. The increasing presence of China in Laos today: a report on fixed point observation of local newspapers from March 2007 to February 2009. *Bulletin of Ritsumeikan Center for Asia Pacific Studies* February.

Gordon, Graham. 2002. *Advocacy toolkit: Practical Action In Advocacy.* Roots Resources 2. Tearfund.   http://tilz.tearfund.org/Publications/ROOTS/Advocacy+toolkit.htm. Accessed 23 April 2013. Original version no longer available online; updated version at http://tilz.tearfund.org/~/media/files/tilz/publications/roots/english/advocacy%20 toolkit/second%20edition/tearfundadvocacytoolkit.pdf.

Hayton, Bill. 2010. Breaking fences in Vietnam. *Forbes*, 13 July.

Kerkvliet, Benedict J. Tria. 2001. An approach for analysing state–society relations in Vietnam. *Sojourn: Journal of Social Issues in Southeast Asia* 16 (2): 238–278.

Koh, Harold Hongju. 1998. 1998 Frankel Lecture: Bringing international law home. *Houston Law Review* 35.

Law on Media. No. 12/1999/QH10 of 12 June 1999. Vietnam.

Law on the Press. 1995. Cambodia.

LICADHO. 2008. *Reading Between the Lines: How Politics, Money & Fear Control Cambodia's Media. Cambodian League for the Promotion and Defense of Human Rights (LICADHO)*. www.licadho-cambodia.org/reports/files/119LICADHOMediaReport2008.pdf. Accessed 12 August 2013.

—— 2010. *Freedom of Expression in Cambodia: The Illusion of Democracy. Cambodian League for the Promotion and Defense of Human Rights (LICADHO)*. www.licadho-cambodia.org/ reports/files/148LICADHOIllusionDemocracy2010.pdf. Accessed 11 October 2012.

Matsumoto, Satoru. 1999. Development and environment in press: a North–South perspective on representations of Mekong hydropower development. Master's thesis, School of Geosciences. Division of Geography, University of Sydney.

McKinley, Catherine. 2011. Vietnam. In *Financially Viable Media in Emerging and Developing Markets*. Paris: World Association of Newspapers and News Publishers.

McQuail, Denis. 1987. *Mass Communication Theory: An Introduction*, 2nd edn. London: Sage.

NGO Forum on Cambodia. 2011. Minute: National Conference on Climate Change, Agriculture and Energy, at Phnom Penh (1–2 December 2011).

Norris, Pippa, and Ronald Inglehart. 2010. Limits on press freedom and regime support. In *Public Sentinel: News Media And Governance Reform*, edited by P. Norris. Washington, DC: World Bank.

Penal Code. 2009. Cambodia.

Penal Code. No. 15/1999/QH10. 21 December 1999. 1999. Vietnam.

Rieu-Clarke, Alistair, and Andrew Allan. 2008. *Role of Water Law: Assessing Governance in the Context of IWRM – An Analysis of Commitment and Implementation in the Tagus and Sesan River Basins*. STRIVER Report No. 6.3. http://kvina.niva.no/striver/Portals/0/ documents/STRIVER_D6_3.pdf. Accessed 3 May 2014.

Roskin, Michael G., Robert L. Cord, James A. Medeiros, and Walter S. Jones. 2008. *Political Science: An Introduction*. Upper Saddle River, NJ: Pearson Prentice Hall.

Samnang, Ham. 2002. Cambodian media in a post-socialist situation. In *Media Fortunes, Changing Times: ASEAN States in Transition*, edited by R. H. K. Heng. Singapore: Institute of Southeast Asian Studies.

Scheufele, D. A. 1999. Framing as a theory of media effects. *Journal of Communication* 49 (1): 103–122.

Seiff, Abby, and Khy Sovuthy. 2012. Sonando said to be 'prisoner of conscience'. *The Cambodia Daily* 3 August.

Simon, Adam F., and Jennifer Jerit. 2007. Toward a theory relating political discourse, media, and public opinion. *Journal of Communication* 57 (2): 254–271.

Subedi, Surya P. 2011. The UN human rights mandate in Cambodia: the challenge of a country in transition and the experience of the special rapporteur for the country. *International Journal of Human Rights* 15 (2): 249–264.

Thayer, C.A. 1982. Laos and Vietnam: the anatomy of a special relationship. In *Contemporary Laos: Studies in the Politics and Society of the Lao PDR*, edited by M. Stuart-Fox. London: University of Queensland Press.

—— 2008. One party rule and the challenge of civil society in Vietnam. In *Remaking the Vietnamese State: Implications for Vietnam and the Region*, Vietnam Workshop, Hong Kong, 21–22 August 2008.

—— 2012. Japan Focus. *Asia-Pacific Journal* 10 (34): 4.

Un, Kheang. 2011. Cambodia: Moving away from democracy? *International Political Science Review/Revue internationale de science politique* 32 (5): 546–562.

UNESCO Office in Phnom Penh. undated. *Literacy*. www.unesco.org/new/en/phnompenh/education/learning-throughout-life/literacy. Accessed 27 November 2013.

United Nations Vietnam. 2010. Basic statistics. www.un.org.vn/en/about-viet-nam/basic-statistics.html. Accessed 14 May 2013.

V1. 2011. Personal interview, 10 November 2011.

—— 2012. Personal interview, 9 July 2012.

V2. 2012. Personal interview, 10 July 2012.

V9. 2012. Personal interview, 13 July 2012.

V10. 2012. Personal interview, 14 July 2012.

V11. 2012. Personal interview, 14 July 2012.

V16. 2011. Personal interview, 23 November 2011.

—— 2012. Personal interview, 19 July 2012.

V19. 2012. Personal interview, 14 August 2012.

Van, Nghia Nguyen. 2011. *The Disappearance of Flooding Season Part 1*. www.youtube.com/watch?v=RT2tTzzQGJQ. Accessed 14 February 2013.

Wagstaff, Jeremy. 2010. *Southeast Asian Media: Patterns of Production and Consumption*. Open Society Foundations. www.opensocietyfoundations.org/publications/southeast-asian-media-patterns-production-and-consumption. Accessed 27 November 2013.

Women's Media Centre of Cambodia. 2012. FAQ. www.wmc.org.kh/page/8. Accessed 12 August 2013.

12 Conclusion

Introduction

This book aims to understand the linkages between rules, norms, and advocacy strategies of NGO coalitions, using the examples of coalitions in Vietnam and Cambodia. In order to understand these linkages, an analytical framework was developed based on the institutional analysis and development (IAD) framework (Ostrom 1999). Comparative case studies were conducted on Cambodian and Vietnamese NGO coalitions and their advocacy work associated with the Xayaburi hydropower dam on the Mekong River in Laos. By comparing two coalitions advocating over the same hydropower dam, the comparative analysis highlighted how rules and norms influenced NGOs' strategies.

This chapter discusses key findings from the analysis of the case studies, and this book's contribution to the body of knowledge on NGO strategies. The book concludes with recommendations for NGO and civil society actors, public decision-makers, and international donors aiming to enhance the role of NGOs and civil society actors in the Mekong region. Recommendations are presented through an analysis of the opportunities and barriers facing NGO coalitions.

Patterns of interactions between formal and informal rules and norms

Through a comparative analysis of how rules and norms influenced four types of strategy adopted by the Rivers Coalition in Cambodia (RCC) and the Vietnam Rivers Network (VRN), the recurring patterns of influence identified were through interactions, namely those among formal rules, informal rules and norms, actors, and biophysical and material conditions. Several types of interaction were observed, described in more detail below.

Complementary interactions

One type of interaction was complementary relationships between formal rules, and between formal rules, informal rules and norms. An example of a complementary relationship was the situation that shaped the VRN's strategies for using science. As discussed in Chapter 6, the formal rules governing the establishment of Vietnamese NGOs provided a platform for NGOs to be closely connected with the Vietnam Union of Science and Technology Associations (VUSTA) (*Decree No. 35/HTBT* 1992; *Decree No. 81/2002/ND-CP* 2002; *Decree No. 88/2003/ND-CP* 2003; *Decree No.30/2012/ND-CP* 2012; *VUSTA Charter* 2012). Another formal rule, Prime Minister's Decision 22/2002/QD-TTg, defines VUSTA's role in reviewing the Vietnamese government's decisions from a scientific perspective (*Decision No. 22/2002/QD-TTg* 2002). This mandate created an opportunity for the VRN to provide a contribution to VUSTA's scientific input to the government on the Xayaburi dam. This complementary relationship between formal rules provided the VRN with an entry point for national decision-makers. In addition, the existence of informal rules and norms in Vietnam that value science interacted with these formal rules, and created situations where the use of science was an effective tool in advocating with decision-makers.

Competing interactions

Another type of interaction was competing relationships between formal rules, as well as between formal rules, informal rules, and norms. An example of this relationship is illustrated in Chapter 11, which examines rules and norms associated with the media. In both Vietnam and Cambodia, formal rules exist that support media freedom (*Constitution of Vietnam* 2001; *Constitution of Cambodia* 1993). However, in reality the media face restrictions through informal rules. In Vietnam, government agencies conduct regular meetings with editors-in-chief about what can and cannot be reported in the media (V9 2012; V19 2012). In Cambodia, most TV and radio stations are controlled by the ruling party (CCIM 2013). A small number of stations run by opposition parties face harassment (CCIM 2013; Freedom House 2012; Un 2011). Competing relationships exist within formal rules that support restriction of the media. In Cambodia, if journalists are too critical of the government they will be prosecuted for disinformation or defamation (C25 2012; LICADHO 2010). In Vietnam, if the media do not follow the 'party line' then editors-in-chief could be prosecuted through the penal code and the media law based on infringement of interests of the state (Article 258 *Penal Code, Vietnam* 1999; Freedom House 2011; *Law on Media* 1999). These examples of competing relationships indicate the ways in which formal systems can be manipulated by the Cambodian and Vietnamese governments when they feel threatened. This situation, as a result, influences the strategies that NGOs can adopt.

What is observed in Cambodia is another example of competing interactions of formal rules, informal rules, and norms. As discussed in Chapter 10, many of the RCC's activities targeting communities facing potential impacts from the Xayaburi dam received informal pressure not to speak up against the authorities (C5 2012; C10 2012; C12 2012; C13 2012). This informal rule, which is a prevailing pressure within Cambodian society, contradicts freedom of expression, the right guaranteed to every Cambodian by the Constitution, one of the formal rules in Cambodia (*Constitution of Cambodia* 1993). The pressure is exerted by the ruling party, a formal organization which dominates the Cambodian political landscape. As discussed in detail in Chapter 10, both local authorities and communities feel pressured to accept a *status quo* due to neo-patrimonialism, an informal way of working in Cambodia which builds on Cambodian traditional culture (Pak *et al.* 2007: 63). Traditional norms of 'respect' and 'fear' towards people who have gained a higher social status, and the teaching of Theravada Buddhism which believes in *karma*, each contribute to the situation where people are reticent to speak out against authorities (Öjendal and Sedara 2006; Rotha and Vannarith 2008: 7; Pak *et al.* 2007: 42). The political system created by formal rules, which places importance on local-level governance, has created a situation where it is critical to win local-level elections in order to sustain political life at the national level (Öjendal and Sedara 2006: 510). This situation motivates the ruling party to attempt winning commune and *sangkat* council elections through the use of neo-patrimonialism. The decentralized political system, which was designed to encourage local participation, effectively counteracts its very purpose through competing interactions between formal rules, informal rules, and norms. These interactions of various formal rules, informal rules, and norms influenced how the RCC could formulate its advocacy strategies.

Role of actors in interactions of rules and norms

These analyses also illustrate that in all the interactions, actors played important roles. This leads to the insight that interactions among rules and norms occur not by themselves, but through actors' adoption and the use of rules and norms. The nature of interaction is therefore dependent on how actors interpret and implement rules and norms.

The analysis of competing relationships between formal rules, informal rules, and norms leads us to consider the role of political power. The analysis highlights that even when formal rules exist to protect rights for civil society actors, the dominant actors can manipulate application of the formal rules through use of informal rules in a way that enhances their political power. In Cambodia and in Vietnam, political power is concentrated in a single political party: the Cambodian People's Party (CPP) and the Communist Party of Vietnam (CPV), respectively. Informal rules, at times, are also used to supplement these political powers.

Actors also play roles in interpreting formal rules in certain ways, and this in turn influences the strategies of the NGO coalitions. The Procedures for Notification, Prior Consultation and Agreement (PNPCA) process of the Xayaburi hydropower dam illustrated the fact that ambiguities in the requirements of the 1995 Mekong Agreement and the PNPCA created the situation where Laos clearly considered the consultation process to be over, and that legitimacy in moving forward with the Xayaburi dam was achieved; while Cambodia and

Vietnam stressed the need for further studies before the construction could move ahead. As a result, the Mekong River Commission (MRC) member states were unable to confirm whether the PNPCA process was complete, the dam construction started without a clear agreement by the member states, and public engagement and information disclosure were conducted only on a limited scale.

Interactions among actors

Another important factor that influenced advocacy strategies was the interactions among the actors themselves, observed in many of the cases. The analysis in Chapter 8 illustrates that the VRN's and RCC's differences in interactions and relationships with the same regional coalition, the Save the Mekong (STM) Coalition, resulted in differences in how they formulated advocacies targeting regional decision-makers. Chapter 9 highlights the importance of personal relationships, an important tactic in approaching national decision-makers, and an important strategy adopted by the VRN.

The profiles and characteristics of actors were also considered as important factors. The fact that some members of the VRN Mekong task force were former high-ranking officials from the Vietnamese government agencies supported the VRN's interactions and relationships with government officials, facilitating its strategy of direct input. In Cambodia, the existing relationships between RCC member NGOs and communities facilitated a wide range of activities undertaken in order to engage local-level stakeholders.

Resource dependence was another important factor that determined actors' relationships and interactions. The relationship between donor and NGO networks (discussed in Chapter 10), and relationships between the case-study NGO networks and the regional STM Coalition (discussed in Chapter 8) are some cases representing the resource-dependent nature of the relationship.

Biophysical and material conditions

Biophysical and material conditions played roles in determining actors' relationships and attitudes towards advocacy targets and issues. In both Cambodia and Vietnam, national government's plans for hydropower dams were important factors that affected the sensitivity of raising the issue with the authorities. For Vietnam, sensitivity was created through international relationships among states, namely between China, Laos, and Vietnam. This became a sensitive issue due to factors such as territorial disputes among the states.

As discussed in Chapter 2, rivers have the nature of common-pool resources, where subtractability of use is high and it is difficult to exclude potential beneficiaries (for a discussion of subtractability see Table 2.4). Therefore, when negotiating over hydropower dams on a single river basin, state actors who have vested interests in building these hydropower dams would naturally need to take into consideration the potential use by other states and its implications.

The case study highlights the importance of taking the nature of such resources into consideration.

Contribution to the body of knowledge

This book contributes to the existing body of knowledge in four ways. First, the study builds on and advances existing studies of formal and informal institutions. As discussed in Chapter 2, Helmke and Levitsky (2004) developed four typologies of relationships between formal and informal institutions. This book's analysis builds on this typology by identifying the nature of interactions between formal rules, informal rules, and norms (Helmke and Levitsky 2004: 728). The book advances the knowledge provided by existing studies as it further analyses how the interactions of formal rules, informal rules, and norms influence the advocacy strategies of NGO coalitions. This is an area identified by Helmke and Levitsky (2004) as in need of further study, and this book provides insights into 'how informal rules shape formal institutional outcomes' (Helmke and Levitsky 2004: 734).

Second, this book advances the study of NGOs, particularly in improving understanding of the factors that influence them. While some existing studies provide insights into how rules and norms influence NGOs' behaviours (Doh and Guay 2006; Ho and Edmonds 2008; Bryant 2001; Brinkerhoff 1999), these studies do not address issues of the influence on NGOs' behaviour of the interactions between formal rules, informal rules, and norms. In addition, the review of the literature (Chapter 2) identified a gap in studies that analyse how rules and norms at different scales – international, national, local, and within-NGO networks – influence NGOs. This book examines how rules and norms at the Mekong regional level, national levels, local levels, and NGO coalition levels affected the advocacy strategies of the NGO coalitions.

Third, the book advances current studies of NGOs in the Mekong region. Many studies of NGOs and civil society in the region have focused on Thailand (Lertchoosakul 2003; Foran 2006). There are also studies on civil society's engagement in water governance in Cambodia and Vietnam (Ha 2011; Trandem 2008). However, comparative analyses of NGOs operating within Cambodia and Vietnam are limited. A recently published study compiling studies of civil society in Cambodia and Vietnam examines differences in state–society relationships, analyses of political space, and tendencies in civil society activities which emerge from political and cultural contexts (Waibel *et al.* 2014). While these existing studies focus on understanding civil society tactics, this book provides additional knowledge by comparing the influence of rules and norms on NGOs in a comparison of Cambodia and Vietnam. The book compares the strategies of two NGO coalitions working towards the same advocacy target in two different contexts.

Finally, and most importantly, this book has developed an analytical framework for analysing the influence of rules and norms on NGO strategies, and has tested the framework by applying it through the analysis of NGO coalitions in Cambodia and Vietnam. As discussed in Chapter 3, the framework was developed as a modification of the IAD framework. The empirical analysis using the analytical framework suggests that this modification was a useful advancement for analysing the influence of rules and norms, in two ways.

First, as noted above, a key finding of this book is that influence occurs through the interactions of different factors – formal rules, informal rules and norms, biophysical and material conditions, and actors. The empirical findings of this research indicate that this interaction occurs in different ways and in combinations of different factors. Compared with the IAD framework, which involves a one-way explanatory direction (how rules influence actors' behaviours), the analytical framework developed for this book allows the analysis of interactions in different directions (how rules influence actors and *vice versa*).

Second, the IAD framework does not distinguish between formal rules and informal rules and norms. Within the IAD framework, all rules are integrated into a single factor, 'rules-in-use'. While Ostrom develops a classification of rules that affect actors' behaviour (Ostrom 2005: 186), she does not specifically distinguish between the formal and informal nature of the rules. Focusing on the delineation between formal rules and informal rules and norms is useful as it provides a research angle of interaction between these two types of rule: one that has clear pathways for modification (formal rules), and the other which does not (informal rules and norms).

The analytical framework also has practical use for NGOs operating within a variety of social contexts. While identification of informal rules and norms is often more challenging than that of formal rules (Helmke and Levitsky 2004: 733), it is important and useful for NGOs to have access to an analytical tool to identify the factors influencing them. Application of the results from academic studies to a real-life operational context is often a challenge. However, the simple analytical framework provided in this book could further facilitate the work of NGOs and civil society actors, particularly those operating in societies with complex sets of rules and norms.

Barriers and opportunities

This book reviews how rules and norms influenced the advocacy strategies of NGO coalitions through a comparative analysis of the advocacy strategies of the RCC and the VRN concerning the Xayaburi hydropower dam. The study provides insights into how different types of rules and norms interacted and were adopted and utilized by actors, and how these affected the two NGO coalitions in their strategy development. The study identifies several barriers and opportunities that NGO and civil society actors face within the context of the Mekong.

Barriers are identified at local, national, and regional levels. In both the Cambodian and the Vietnamese contexts, contradictory relationships between formal rules and informal rules and norms created barriers to implementing some of the strategies of the case-study NGO coalitions. The strategies affected by these contradictions include the use of media, and awareness-raising activities with communities. In both activities, the influence of informal rules and norms was stronger than that of formal rules, creating a competing relationship. While formal rules typically have formal pathways to modify them, informal rules and norms usually do not have formal ways of modification, thus they are harder to change (Williamson 2000: 579). NGO coalitions working in this context therefore had to find a way to minimize the negative impacts of the influence of informal rules and norms. When informal rules are exercised by powerful individuals or organizations within a society, it is often difficult to avoid this impact.

The close relationship between the state and NGOs in Vietnam could at times create a situation where it was difficult for the NGOs to be critical of the government. This was illustrated by one comment from an interviewee, who noted that during a workshop which the VRN organized with VUSTA on the Xayaburi hydropower dam, guest speakers were allowed to discuss dams on the mainstream of the Mekong, but not on the tributaries (R8 2012). This is associated with the fact that while Vietnam is critical of impacts from the Xayaburi dam, it has, at the same time, already built several dams on the tributaries of the Mekong which have impacted downstream communities, and continues to build more dams on other tributaries of the Mekong.

Another barrier faced by both case-study NGO coalitions, as well as other NGOs working within the region, was the lack of pathways to engage and influence decision-making over development issues associated with an international river. Integration of some of the existing international rules and norms by the MRC member states could provide opportunities for NGO and civil society actors to enhance their roles in the governance of the Mekong River. For instance, the Aarhus Convention allows the public to

> participate in decision-making and to have access to justice in environmental matters without discrimination as to citizenship, nationality or domicile, and in the case of a legal person, without discrimination as to where it has its registered seat or an effective centre of its activities.
>
> (Article 3 (9) *Aarhus Convention* 1998)

A similar provision is available in the 1997 *UN Watercourses Convention* Article 32, which indicates that the watercourse state shall provide access to judicial or other procedures, and the right to claim compensation or other relief for individuals or organizations who suffer or may potentially suffer significant transboundary harm as a result of activities related to an international watercourse (Article 32 *UN Watercourses Convention* 1997). Article 2 of the Convention on Environmental Impact Assessment in a Transboundary Context (Espoo Convention) requires the contracting state to establish a procedure permitting public participation in transboundary environmental impact assessment (*Espoo Convention* 1991). Referring to international norms associated with the human right to water, the United Nations Committee on Economic, Social and Cultural Rights suggests that

> The right of individuals and groups to participate in decision-making process that may affect their exercise of the right to water must be an integral part of any policy, programme, or strategy concerning water.
> (United Nations Economic and Social Council 2003: para 48; Cullet and Gowlland-Gualtieri 2005: 311)

Specifically associated with the development of dams, the World Commission on Dams made a statement that free prior and informed consent (FPIC) principles should guide the building of dams that may affect indigenous people and ethnic minorities (Tamang 2005: 3; Cullet and Gowlland-Gualtieri 2005: 310).

The case studies also identified areas of opportunity in enhancing the advocacy strategies of NGO coalitions within national domains. For example, the collaboration between the VRN and VUSTA was enhanced through formal rules that defined their relationships. The strategy to use science targeted at national decision-makers was also possible through complementary interactions between formal rules, informal rules and norms, actors, and biophysical and material conditions. In Cambodia, an opportunity for collaboration with the government was created through the PNPCA process, a formal rule within the MRC member countries.

These windows of opportunity could be examined in future studies through analyses of the interactions of the key elements of the analytical framework developed for this book, including formal rules, informal rules and norms, actors, and biophysical and material conditions. While situations associated with actors' positions can shift over time, the analytical framework presented in this book could provide a useful tool for NGO actors who need to develop advocacy strategies.

Areas for future study

One potential area for future studies is analysis of the influence of rules and norms on the strategies of NGOs and civil society actors operating in other geographical areas. This book compares case studies of NGO coalitions in Cambodia and Vietnam. Additional studies reviewing NGOs and civil society actors operating in other countries within the Mekong River Basin, as well as actors operating at the whole-basin scale, would provide insights into how various rules and norms influence different types of NGOs and civil society actors within the region. Such a study would have both academic and practical benefit. Academically, it could be an opportunity to test the applicability of the analytical framework at the multi-national scale. It could also contribute to the body of knowledge in NGO studies within the Mekong region. Practically, such studies could provide a basis for suggesting improvements in the formal rules existing within the Mekong River Basin, which would further enhance the role of NGOs and civil society actors in the governance of transboundary rivers. Similar studies applying this book's approach and analytical framework to other river basins could also advance the methodology used here.

Another suggested area of future study is to conduct deeper analyses into specific aspects of the book's findings. One such study could involve understanding the long-term implications of the interactions between rules, norms, and actors, reflecting the feedback arrow connecting 'outcomes' and 'exogenous variables' in the original IAD framework. As discussed in Chapter 3, this aspect is beyond the scope of this book and is thus excluded from the analytical framework adopted here. However, additional studies that consider the long-term implications of interactions among rules, norms, and actors, and the resultant strategies adopted by NGO actors, may reveal the mechanism of how rules and norms may change as a result of these interactions. Further elaboration of the feedback arrow from strategies to rules and norms illustrated in this book's analytical framework may benefit in understanding such long-term implications.

Following up on the finding of the importance of interactions among actors (discussed above), long-term research on how rules and norms influence NGOs could benefit from reflecting on key aspects of the advocacy coalition framework proposed by Sabatier and Jenkins-Smith (1999). As discussed in Chapter 2, the advocacy coalition framework is used to examine the process and development of advocacy coalitions that consist of people from governmental and private organizations who share a set of normative beliefs and engage in a certain level of coordinated activity over time (Sabatier and Jenkins-Smith 1999: 120). In relation to the case-study coalitions in this book, the VRN in particular has the characteristic of an advocacy coalition, as its members include both NGO members and individual members from a wide variety of affiliations, including government. The advocacy coalition framework provides a hypothesis that allies and opponents of coalitions tend to become stable over a decade (Sabatier and Jenkins-Smith 1999: 124). Therefore understanding the dynamics of relationships among actors within coalitions preferably needs to be examined over a decade or longer.

Another key finding that could benefit from follow-up research is the role of actors in determining the types of interaction between formal and informal rules and norms. As discussed above, this aspect is reflective of the role of political power. Further analysis of how power dynamics could determine the interactions among rules, norms, and actors could benefit from existing studies of power such as that of Lukes (2005), who provides three dimensions of power; and that of Gaventa (2005), who proposes use of a 'power cube' to analyse the dynamics of civil society participation and engagement.

Further analyses conducted alongside the results in this book should assist in advancing the role of civil society actors in the governance of transboundary water.

References

1992 Constitution of the Socialist Republic of Vietnam: As Amended 25 December 2001 (transl. Allens Arthur Robinson). 2001. Vietnam.

Brinkerhoff, Derick W. 1999. State–civil society networks for policy implementation in developing countries. *Review of Policy Research* 16 (1): 123–147.

Bryant, Raymond L. 2001. Explaining state–environmental NGO relations in the Philippines and Indonesia. *Singapore Journal of Tropical Geography* 22 (1): 15–37.

C5. 2012. Personal interview, 25 July 2012.

C10. 2012. Personal interview, 28 July 2012.

C12. 2012. Personal interview, 30 July 2012.

C13. 2012. Personal interview, 31 July 2012.

C25. 2012. Personal interview, 8 August 2012.

CCIM. 2013. *Challenges for Independent Media Development in Cambodia.* Phnom Penh: Cambodian Center for Independent Media. www.ccimcambodia.org/report/CCIM_report_indepedent_media_promotion.pdf. Accessed 11 August 2013.

Charter: The Vietnam Union of Science and Technology Associations. 2012. Vietnam.

The Constitution of the Kingdom of Cambodia. 1993. Cambodia.

Convention on Environmental Impact Assessment in a Transboundary Context (Espoo Convention). 1991. Adopted on 25 February 1991, entered into force on 10 September 1997.

Convention on the Law of the Non-navigational Uses of International Watercourses. 1997. Adopted on 21 May 1997, entered into force on 17 August 2014. Reprinted in (1997) 36 ILM 700.

Cullet, Philippe, and Alix Gowlland-Gualtieri. 2005. Local communities and water investments. In *Fresh Water and International Economic Law*, edited by E. B. Weiss, L. B. De Chazournes, and N. Bernasconi-Osterwalder. Oxford: Oxford University Press.

Decision 22/2002/QD-TTg of 30 January 2002 of the Prime Minister on the Activities of Consultancy, Judgment and Social Expertise by Vietnam Union of Scientific and Technical Associations 2002. Vietnam.

Decree 35/HTBT of 28 January 1992 of the Council of Ministers on Establishment of Non-profit and Science and Technology Organization. 1992. Vietnam.

Decree No. 30/2012/ND-CP of 12 April 2012 of the Government on the Organization and Operation of Social Funds and Charity Funds. 2012. Vietnam.

Decree No. 81/2002/ND-CP of 17 October 2002 of the Government on Detailing the Implementation of a Number of Articles of the Science and Technology Law. 2002. Vietnam.

Decree No. 88/2003/ND-CP of 30 July 2003 of the Government on Providing for the Organization, Operation and Management of Associations. 2003. Vietnam.

Doh, Jonathan P. and Terrence R. Guay. 2006. Corporate social responsibility, public policy, and NGO activism in Europe and the United States: an institutional–stakeholder perspective. *Journal of Management Studies* 43 (1): 47–73.

Foran, Tira. 2006. Rivers of contention: Pak Mun Dam, electricity planning, and state-society relations in Thailand 1932–2004. PhD thesis, Division of Geography, School of Geosciences, University of Sydney.

Freedom House. 2011. *Vietnam: Freedom of Press 2011.* www.freedomhouse.org/report/freedom-press/2011/vietnam. Accessed 26 November 2012.

—— 2012. *Freedom of the Press 2012 – Cambodia.* www.unhcr.org/refworld/country,,,,KHM,,507bcae6c,0.html. Accessed 22 October 2012.

Gaventa, John. 2005. *Reflections on the Uses of the 'Power Cube' Approach for Analyzing the Spaces, Places and Dynamics of Civil Society Participation and Engagement*. CFP Evaluation Series 2003–2006, No 4. The Netherlands: Mfp Breed Netwerk.

Ha, Tran Van. 2011. Local people's participation in involuntary resettlement in Vietnam: a case study of Son La hydropower project. In *Water Rights and Social Justice in the Mekong Region*, edited by K. Lazarus, N. Badenoch, N. Dao, and B. P. Resurreccion. London: Earthscan.

Helmke, G., and S. Levitsky. 2004. Informal institutions and comparative politics: a research agenda. *Perspectives on Politics* 2 (4): 725–740.

Ho, P., and R. L. Edmonds, eds. 2008. *China's Embedded Activism: Opportunities and Constraints of a Social Movement*. London: Psychology Press.

Law on Media. No. 12/1999/QH10 of 12 June 1999. Vietnam.

Lertchoosakul, K. 2003. Conceptualising the roles and limitations of NGOs in the anti-Pak Mun dam movement. In *Radicalising Thailand: New Political Perspectives*, edited by J. Ungpakorn. Bangkok: Institute of Asian Studies, Chulalongkorn University.

LICADHO. 2010. *Freedom of Expression in Cambodia: The Illusion of Democracy*. Cambodian League for the Promotion and Defense of Human Rights (LICADHO). www.licadho-cambodia.org/reports/files/148LICADHOIllusionDemocracy2010.pdf. Accessed 11 October 2012.

Lukes, Steven. 2005. *Power: A Radical View*, 2nd edn. New York: Palgrave Macmillan.

Öjendal, Joakim, and Kim Sedara. 2006. *Korob, Kaud, Klach*: in search of agency in rural Cambodia. *Journal of Southeast Asian Studies* 37 (3): 507.

Ostrom, Elinor. 1999. Institutional rational choice: an assessment of the institutional analysis and development framework. In *Theories of the Policy Process*, edited by P. A. Sabatier. Oxford: Westview Press.

—— 2005. *Understanding Institutional Diversity*. Princeton, NJ: Princeton University Press.

Pak, K., V. Horng, N. Eng, S. Ann, S. Kim, J. Knowles, and D. Craig. 2007. *Critical Literature Review on Accountability and Neo-Patrimonialism: Theoretical Discussions and the Case of Cambodia*. Working Paper 34. Phnom Penh: Cambodia Development Resource Institute.

Penal Code. No. 15/1999/QH10. 21 December 1999. 1999. Vietnam.

R8. 2012. Personal interview, 14 August 2012.

Rotha, Chan, and Chheang Vannarith. 2008. Cultural challenges to the decentralization process in Cambodia. *Ritsumeikan Journal of Asia Pacific Studies* 24: 1–16.

Sabatier, Paul A., and Hank C. Jenkins-Smith. 1999. The advocacy coalition framework: an assessment. In *Theories of Policy Process*, edited by P. A. Sabatier. Oxford: Westview Press.

Tamang, Parshuram. 2005. An overview of the principle of free, prior and informed consent and indigenous peoples in international and domestic law and practices. In *Workshop on Free, Prior and Informed Consent*, New York, 17–19 January 2005. New York: United Nations Department of Economic and Social Affairs.

Trandem, Ame. 2008. A Vietnamese/Cambodian transboundary dialogue: impacts of dams on the Se San river. *Development* 51 (1): 108–113.

Un, Kheang. 2011. Cambodia: moving away from democracy? *International Political Science Review/Revue internationale de science politique* 32 (5): 546–562.

UNECE Convention on Access to Information, Public Participation in Decision-making and Access to Justice in Environmental Matters (Aarhus Convention). 1998. Adopted on 25 June 1998, entered into force on 30 October 2001. 2161 UNTS 447; (1999) 38 ILM 517.

United Nations Economic and Social Council. 2003. Substantive issues arising in the implementation of the international covenant on economic, social and cultural rights.

General comment No. 15 (2002). The right to water (arts. 11 and 12 of the International Covenant on Economic, Social and Cultural Rights). In *Committee on Economic, Social and Cultural Rights. Twenty-ninth Session*, Geneva, 11–29 November 2002. New York: United Nations Economic and Social Council.

V9. 2012. Personal interview, 13 July 2012.

V19. 2012. Personal interview, 14 August 2012.

Waibel, Gabi, Judith Ehlert, and Hart N. Feuer, eds. 2014. *Southeast Asia and the Civil Society Gaze: Scoping a Contested Concept in Cambodia and Vietnam*. London: Routledge.

Williamson, Oliver E. 2000. The new institutional economics: taking stock, looking ahead. *Journal of Economic Literature* 38 (3): 595–613.

Annex: Interviews and meetings

Interviews: Cambodia

Number	Interviewee
C1	RCC member
C2	Non-RCC member informant (NGO)
C3	RCC member
C4	Non-RCC member informant (NGO)
C5	RCC member
C6	RCC member
C7	Non-RCC member informant
C8	Non-RCC member informant (NGO)
C9	RCC member
C10	Non-RCC member informant (NGO)
C11	Non-RCC member informant (NGO)
C12	RCC member
C13	RCC member
C14	RCC member
C15	RCC member
C16	RCC member
C17	Non-RCC member informant (community)
C18	RCC member
C19	RCC member
C20	Non-RCC member informant (community)
C21	RCC member
C22	RCC member
C23	RCC member
C24	Non-RCC member informant (government)
C25	Non-RCC member informant (journalist)
C26	Non-RCC member informant (government)

Number	Interviewee
C27	Non-RCC member informant (donor)
C28	Former RCC member
C29	Non-RCC member informant (government)
C30	Non-RCC member informant (NGO)
C31	Non-RCC member informant (community)
C32	Non-RCC member informant (government)
C33	Non-RCC member informant (former government)
C34	Non-RCC member informant (government)
C35	Former RCC member

Interviews: Vietnam

Number	Interviewee
V1	VRN member
V2	VRN member
V3	Non-VRN member (NGO)
V4	Non-VRN member (NGO)
V5	VRN member
V6	Non-VRN member (government)
V7	Non-VRN member
V8	Non-VRN member (NGO)
V9	Non-VRN member (journalist)
V10	VRN member
V11	VRN member
V12	VRN member
V13	VRN member
V14	VRN member
V15	Non-VRN member (NGO)
V16	VRN member
V17	VRN member
V18	Non-VRN member
V19	Non-VRN member (journalist)
V20	Non-VRN member (NGO)
V21	Non-VRN member
V22	Non-VRN member
V23	VRN member
V24	Non-VRN member (government)

Number	Interviewee
V25	VRN member
V26	VRN member

Interviews: regional

Number	Interviewee
R1	Informant from NGO
R2	Informant from NGO
R3	Informant from a university
R4	Informant from inter-governmental organization
R5	Informant from inter-governmental organization
R6	Informant from inter-governmental organization
R7	Informant from a donor agency
R8	Informant from NGO
R9	Informant from NGO
R10	Informant from NGO
R11	Informant from research institute

Workshops and meetings observed

Date of event	Type of workshops/meetings
22 November 2011	Training workshop on Coastal Geomorphology and Sediment Transit. Organized by VRN and WWF. Ho Chi Minh City, Vietnam.
23 November 2011	Xayaburi hydropower project – gains and losses for the Mekong River Basin. Organized by VUSTA and VRN.
24 November 2011	Presentation and discussion: Snapshot of Civil Society in Vietnam. Organized by People's Participation Working Group. Hanoi, Vietnam.
1–2 December 2011	First Mekong Resources Forum. Water Resources and Sustainable Development: Perspectives from Laos and Vietnam. Organized by Pan Nature. Hanoi, Vietnam.
29 June–2 July 2012	Mekong Legal Network 4 Regional Meeting. Chiang Mai, Thailand.
20–21 July 2012	10 Wetland training course. Organized by university network for wetland research and training in the Mekong region. An Giang University, Vietnam.
14 August 2012	Seminar: Mekong and the Hydropower Dam. Organized by VUSTA and VRN. Ho Chi Minh City, Vietnam.
16–17 August 2012	RCC network meeting. Phnom Penh, Cambodia.

Index

For Product Safety Concerns and Information please contact our EU
representative GPSR@taylorandfrancis.com
Taylor & Francis Verlag GmbH, Kaufingerstraße 24, 80331 München, Germany

www.ingramcontent.com/pod-product-compliance
Lightning Source LLC
Chambersburg PA
CBHW050421280326
41932CB00013BA/1950

9 780815 395379